T0155553

SpringerBriefs in Applied Sciences and Technology

PoliMI SpringerBriefs

Editorial Board Members

Barbara Pernici, Politecnico di Milano, Milano, Italy
Stefano Della Torre, Politecnico di Milano, Milano, Italy
Bianca M. Colosimo, Politecnico di Milano, Milano, Italy
Tiziano Faravelli, Politecnico di Milano, Milano, Italy
Roberto Paolucci, Politecnico di Milano, Milano, Italy
Silvia Piardi, Politecnico di Milano, Milano, Italy

More information about this series at http://www.springer.com/series/11159
http://www.polimi.it

Fabio Borghetti · Paolo Cerean ·
Marco Derudi · Alessio Frassoldati

Road Tunnels

An Analytical Model for Risk Analysis

POLITECNICO
MILANO 1863

 Springer

Fabio Borghetti
Department of Design
Politecnico di Milano
Milan, Italy

Paolo Cerean
Department of Design
Politecnico di Milano
Milan, Italy

Marco Derudi
Department of Chemistry, Materials
and Chemical Engineering "Giulio Natta"
Politecnico di Milano
Milan, Italy

Alessio Frassoldati
Department of Chemistry, Materials
and Chemical Engineering "Giulio Natta"
Politecnico di Milano
Milan, Italy

ISSN 2191-530X   ISSN 2191-5318 (electronic)
SpringerBriefs in Applied Sciences and Technology
ISSN 2282-2577   ISSN 2282-2585 (electronic)
PoliMI SpringerBriefs
ISBN 978-3-030-00568-9   ISBN 978-3-030-00569-6 (eBook)
https://doi.org/10.1007/978-3-030-00569-6

Library of Congress Control Number: 2019933204

© The Author(s), under exclusive licence to Springer Nature Switzerland AG 2019
This work is subject to copyright. All rights are solely and exclusively licensed by the Publisher, whether the whole or part of the material is concerned, specifically the rights of translation, reprinting, reuse of illustrations, recitation, broadcasting, reproduction on microfilms or in any other physical way, and transmission or information storage and retrieval, electronic adaptation, computer software, or by similar or dissimilar methodology now known or hereafter developed.
The use of general descriptive names, registered names, trademarks, service marks, etc. in this publication does not imply, even in the absence of a specific statement, that such names are exempt from the relevant protective laws and regulations and therefore free for general use.
The publisher, the authors and the editors are safe to assume that the advice and information in this book are believed to be true and accurate at the date of publication. Neither the publisher nor the authors or the editors give a warranty, express or implied, with respect to the material contained herein or for any errors or omissions that may have been made. The publisher remains neutral with regard to jurisdictional claims in published maps and institutional affiliations.

This Springer imprint is published by the registered company Springer Nature Switzerland AG
The registered company address is: Gewerbestrasse 11, 6330 Cham, Switzerland

Preface

Road tunnels allow on the one hand to improve the plane-altimetry coordination of road sections, reducing slopes, at times even lengths (distances) and fuel consumption, but on the other hand they can be a serious problem for the safety of users in case of a relevant accidental event.

Some European road tunnels entered into service many years ago: they were designed in a period when knowledge, technical possibilities and transport conditions were very different from those of today. As an example, in some realities the number of heavy vehicles represent even 50% of the total vehicles. Even current traffic volumes are different from those of a few decades ago.

There is also a difference concerning technological innovation which today permits the use of safety systems and devices that were inconceivable until just a few years ago.

In Italy, where more than 50% of the road tunnels belonging to *TERN (Trans European Road Network)* are situated, the panorama is quite varied. There are new tunnels, designed and built during the last decade, but there are also many operative tunnels that were designed and realised a few decades ago.

This could imply different levels of safety, which must be improved and made uniform in order to create a common standard.

At a statistical level, the tunnels are marked by a lower accident rate than the open-air sections because the user is more careful when driving and the weather and visibility conditions are constant.

On the contrary, the possible consequences of a relevant event in a confined place such as a tunnel become significantly more important if fatalities and damage to the infrastructure are considered.

Following the accident in the Mont Blanc tunnel in 1999, which caused 39 fatalities and huge damages to the tunnel structure, the European Community issued Directive 2004/54/EC on the Minimum Safety Requisites for tunnels of the TERN network that are longer than 500 m.

In reality, after the Mont Blanc Tunnel episode, other tragic events occurred in some European tunnels, for example the St. Gotthard tunnel in 2001, the Tauern Tunnel in 2002 and the Fréjus Road Tunnel in 2005, which sensitised public and political opinion to the problem, highlighting the importance of this type of infrastructure at human, economic and cultural levels.

The Directive disciplines tunnel safety, establishing the minimum requisites and identifying risk analysis as the analytical and well-defined method for determining the risk level of every tunnel.

Risks must be evaluated considering the use of measures regarding tunnel geometry and its infrastructures, as well as equipment and management procedures.

The interregional corridors, marked often by the presence of tunnels, are important connections for the mobility of people and goods. This is why elevated standards of safety have to be constantly guaranteed.

A relevant accident in a tunnel affects not only the users (exposed populations) that are inside it, but also has social and economic effects that occur in an area (region or state) following the either short- or long-term closure of the tunnel.

These effects can be connected with the loss of freight and passenger traffic (even tourists) and the increase in travel times caused by traffic deviation to other alternative ways. On the basis of the importance and localisation of a tunnel, it is therefore important to also consider the potential impacts on the socio-economic system of the country.

The tunnel closure due to a relevant event interrupts the whole section, generating a diversion of traffic to alternative, and maybe already congested, routes. In this sense, the tunnel system plays an important role in the evaluation of the systemic vulnerability, so as to determine the possible effect on the whole transport network after a road section has to be closed.

Consequently, knowing the technical characteristics of a tunnel makes it possible to evaluate its resilience following a relevant event, not just with reference to the potential damage for the exposed users, but also to the effects tied to the more or less extended closure period needed to restore normal working conditions.

The purpose of the book is therefore to give a valid contribution to research, proposing an innovative quantitative analysis model for risk that can be applied to road tunnels.

This book is organized into chapters as follows:

- Chapter 1—description and overview of risk analysis and assessment in road tunnels;
- Chapter 2—evaluation of the state of the art regarding the methods and models for tunnel risk assessment;
- Chapter 3—description of the structure of the model proposed, with reference to the event tree analysis for estimating the accidents frequencies and the position of accidental scenarios in tunnels;
- Chapter 4—availability of infrastructure measures, equipment and management procedures for tunnel safety, taking into account their combinations and including the analysis of the respective interdependence;
- Chapter 5—description of the queue formation model of vehicles that are stopped in tunnels with reference to its characteristic parameters;

- Chapter 6—description of the distribution model of potentially involved users;
- Chapter 7—description of the approach used to determine the consequences evolution of the accidental scenarios;
- Chapter 8—description of the user egress model with particular reference to the egress time evaluation;
- Chapter 9—calculation of the F-N curve and EDV (Expected Damage Value) estimation starting from both frequencies of occurrence and consequences of the accidental scenarios;
- Chapter 10—description of model calibration and validation phases, with reference to automation of the calculation process, to sensitivity analysis and to a comparison with a literature software.

Milan, Italy

Fabio Borghetti
Paolo Cerean
Marco Derudi
Alessio Frassoldati

Acknowledgements

The authors wish to thank all those who made the preparation of this book possible. In particular, they thank the company Strada dei Parchi S.p.A., which manages a large number of tunnels belonging to the A24 and A25 motorways in Italy, for believing in this experience.

Contents

Symbols, Acronyms and Abbreviations

Here below the description of the model subscripts:

i Represents the lane of the tunnel;

j Represents the cell into which the queue is divided;

m Represents infrastructure measures and management procedures inside the tunnel;

p Represents the position along the tunnel at which the initial event can occur;

s Represents the accidental scenario that can occur inside the tunnel (event dynamics).

Parameters

$\%BUS_i$	% of bus on total traffic in the i-th lane
$\%HV_i$	% of heavy vehicles that are part of the total traffic in the i-th lane
$\%LV_i$	% of light vehicles that are part of the total traffic in the i-th lane
ADT	Average Daily Traffic
ADR	European Agreement concerning the International Carriage of Dangerous Goods by Road (**A**ccord européen relatif au transport international des marchandises **D**angereuses par **R**oute)
$ALARP$	As Low As Reasonably Practicable
$ANAS$	Azienda Nazionale Autonoma delle Strade, Italian public company that manages a large number of roads and tunnels
$ASET$	Available Safe Egress Time
$BLEVE$	Boiling Liquid Expanding Vapour Explosion
CFD	Computational Fluid.Dynamics

$Cint_m$	Coefficient at the mth intermediate variable between 0 and 1
cm	cth combination of infrastructure measures, equipment and management procedures hypothesising that each measure MI_m, if present can be available or not available
Cpr_m	Coefficient associated with the m-th primary measure. If the measure is present, Cpr_m is worth $Cprpres_m$, otherwise it is worth 0
$Cprpres_m$	Value of Cpr_m if the primary measure is present. The sum of the values of all the $Cprpres_m$ is 1
c_s	Coefficient that considers the effects of the sth accidental scenario on the user movement speed v_m
$Cint_m$	Coefficient associated with the m-th intermediate measure, variable between 0 and 1
$Csec_m$	Coefficient associated with the m-th secondary measure, variable between 1 and $Csecmax_m$
$Csecmax_m$	Maximum value of $Csec_m$ associated with the mth secondary measure
d_{BUS}	Average bus length
$dcell_{i,j}$	Distance between the centre of gravity $G_{i,j}$, of the j-th cell belonging to the i-th lane and the corresponding EE_k emergency exit
$Dqueue_i$	Density of the vehicles stopped in the queue in the i-th lane
DG	Dangerous Goods
$dul_{i,m,p}$	Density of users in the i-th lane
$dvehic$	Average safety distance between the vehicles stopped in the queue
$dx_{i,j}$	Longitudinal component of the distance covered by users associated with the j-th cell belonging to the i-th lane to reach the emergency exit. This distance is considered from the cell centre of gravity of the j-th cell
$dy_{i,j}$	Transversal component of the distance covered by the users associated with the j-th cell belonging to the i-th lane to reach the emergency exit. This distance is considered from the centre of gravity of the j-th cell
EDV	Expected Damage Value given by the integral of the area subtended by the F-N curve
EE_k	k-th emergency exit
$Ffire$	Frequency of occurrence of the initial fire event
$Fpos_{p,s}$	Frequency of occurrence of a specific accidental scenario associated with a position of the tunnel PO_p
$Fpos_{p,s}$ cumulated	Cumulated sum of $Fpos_{p,s}$
$Frel$	Frequency of occurrence of the initial DG release event
F_s	Frequency of occurrence of the sth accidental scenario
GSM	Global System for Mobile Communications

H_{cm}	Probability associated with each combination of infrastructure measures, equipment and management procedures MI_m as a function of their presence and reliability (product of Pry and Prn)
H_{cm} cumulated max	Threshold of cumulated probability H_{cm} that determines the number of combinations of measures c_m to be analysed
I	Total number of lanes in the tunnel
$Jz1_i$	Number of cells present in zone $z1_s$ for the i-th lane
$Jz2_i$	Number of cells present in zone $z2_s$ for the i-th lane
KA_i	Vehicular density in section A of the i-th lane in free flow conditions
KB_i	Vehicular density in section B of the i-th lane in free flow conditions
lc	Length of the j-th cell in which the queue of vehicles is discretized
$Lqcl1_{i,m}$	Length of the queue of vehicle formed between the initial event occurrence and the tunnel closure
$Lqcl2_{i,m}$	Length of the queue formed by the vehicles that have entered the tunnel in addition to the already-stopped vehicles until the tunnel closure
$Lqcltot_{i,m}$	Total length of the queue in the ith lane associated with the closing time tch_m
$Lc_{i,p}$	Section of tunnel between the end of the queue $Lqcl1_{i,m}$ and the tunnel entrance $(L_p - Lqcl1_{i,m})$
$Lqueue_{i,m,p}$	Length of the queue of the ith lane according to the measures and position of the initial event
L_p	Distance between the initial event position and the tunnel entrance
$lBUS$	Average length of the buses
lHV	Average length of the heavy vehicles
lLV	Average length of the light vehicles
$Ltot$	Total tunnel length
M	Number of total infrastructure measures, equipment and management procedures considered by the model
MI_m	mth infrastructure measure, equipment and/or management procedure of the model
$Mint$	Total number of intermediate measures
Mpr	Total number of primary measures
$Msec$	Total number of secondary measures
MPV	Multi-Purpose Vehicle
My	Infrastructure measure, equipment and/or management procedure available in a tunnel among the M considered by the model
N	Total number of expected fatalities associated with a tunnel

$Ncellfatz2_{i,j,m,p,s}$	Number of fatalities associated with the j-th cell located in the ith lane and belonging to $z2_s$
$NCH_{p,s}$	Number of expected fatalities weighted with the probability of the measure combination H_{cm}
$NC_{i,j,m,p}$	Number of users associated with each cell j in which the queue of vehicles is discretized
$NCT_{i,m,p}$	Total number of cells used to discretize the queue of vehicles of the i-th lane
$N_{i,m,p}$	Number of users inside the queued vehicles of the i-th lane that are potentially exposed to the consequences of an accidental event
$N_{m,p,s}$	Total number of fatalities inside the tunnel
$Ntotfatz1_{m,p,s}$	Total number of fatalities in zone 1
$Ntotfatz2_{m,p,s}$	Total number of fatalities in zone 2
$nvehic_i$	Number of vehicles involved by an accidental event in the i-th lane
O	Number of the total PO_p positions considered by the model
$OBUS$	Average coefficient of bus occupation
OHV	Average coefficient of heavy vehicle occupation
OLV	Average coefficient of light vehicle occupation
PO_p	p-th position in which the initial event can occur, expressed as a percentage as to the total length of the tunnel Ltot
$Ppos_p$	Probability of occurrence of the s-th accidental scenario in the p-th position along the tunnel
Prn	Value assumed by RM_m if the measure MI_m is not available
Pry	Value assumed by RM_m if the measure MI_m is available
PEE_k	Position of the k-th emergency exit with respect to the tunnel entrance
QA_i	Vehicular flow in section A of the i-th lane in free flow conditions
QB_i	Vehicular flow in section B of the i-th lane in free flow conditions
RM_m	Parameter of reliability associated with the presence and availability of the m-th infrastructure measure, equipment and/or management procedure
$RSET$	Required Safe Egress Time
S	Total number of accidental scenarios considered by the model
SC_s	s-th accidental scenario
$tclbase$	Minimum time that permits tunnel closing (hypothesising to stop the vehicles near the entrance) following the identification and validation of an initial event (e.g. fire or DG release)

tcl_m	Closing time of a tunnel according to the role of infrastructure measures, equipment and management procedures MI_m. This time is greater than or equal to $tchbase$
$Tevac_{i,j,m,p}$	Total evacuation time of the users in the j-th cell of each lane
$Tfc_{i,j,m,p}$	Queue formation time
Tf_s	Time related to the dynamics of the effects of the s-th accidental scenario
$Tmov_{i,j,m}$	Movement time of the users during the egress process
$Tper_{i,j,m,s}$	User permanence time in the critical zone $z2_s$ during the egress process
$Tpmax$	Tenability threshold in zone $z2_s$ related to the effects of an accidental scenario
$Tppr_{i,j,m,p,s}$	Permanence time of the users of the j-th *cell* of $z2_s$ during the pre-movement phase
$Tpr_{i,j,m}$	Pre-movement time
$Trbase$	Response time following the recognition of an initial event by the users
$Trecbase$	Basic recognition time of an accidental event by the users
$Trec_m$	Recognition time actually required by the users to recognise the occurrence of an initial event
Tr_m	Response time that passes from the recognition of an accidental event and the beginning of the evacuation from the vehicle
$tsat_{i,p}$	Time during which the distance L_p between the initial event and the tunnel entrance fills with vehicles
$TuBUS$	Average time required by users to exit a bus
$TuHV$	Average time required by users to exit a heavy vehicle
$TuLV$	Average time required by users to exit a light vehicle
$Tuvehic_i$	Time required by users to exit a vehicle. This time is given by the weighted average of $TuBUS$, $TuHV$, $TuLV$ according to the composition of the traffic in the i-th *lane*
u_i	Queue formation speed in the i-th lane as a consequence of an accidental event
$vbase$	Basic movement speed of the users from the centre of gravity of the j-th cell to the emergency exit
$vBUS$	Average bus speed
vHV	Average speed of the heavy vehicles
v_i	Flow speed of the i-th lane given by the weighted average of vLV, vHV, and $vBUS$
vFL_i	Speed of the free vehicular flow in the i-th lane
vLV	Average speed of the light vehicles
v_m	User movement speed from the centre of gravity of the j-th cell to the emergency exit
VCE	Vapour Cloud Explosion

VMP	Variable Message Panel
$xG_{i,j}$	Distance between the centre of gravity $G_{i,j}$ of the j-th cell and the initial event
Y	Maximum distance accepted by the users of the j-th cell to reach the emergency exit moving towards the initial event (running direction)
$z1_s$	Tunnel zone in which the users are affected by the lethal effects of the s-th accidental scenario
$z2_s$	Tunnel zone in which the egress and tenability of the users affected by the s-th accidental scenario have to be verified

Chapter 1
Road Tunnels Risk Analysis

Abstract The risk analysis process for road tunnels with particular reference to the concept of risk and safety is described in this chapter. In particular, the specific characteristics of the risk models and the techniques of risk representation and acceptance are presented.

Tunnel risk analysis methods are aimed at evaluating and managing the risk associated with a specific tunnel system in relation to the consequences on the potentially exposed population [1–3].

Risk evaluation is a process that leads to the identification of the possible dangers and to the determination of the risks for the population that is potentially exposed to damages or safety issues which can derive from an accidental event; this also includes the estimation of the uncertainties related to the risk evaluation process.

Risk management is a decision-making process, subsequent to risk evaluation, that involves the realisation of safety measures and/or procedures which have to be harmonized with the social, economic, and political context in which the activity is realised.

The following are the phases of the risk analysis method:

- identification of the possible dangers connected with the tunnel system;
- mapping of the random factors that are potentially responsible for system deviation from the successful trajectory;
- formulation of models that represent the accidents trajectories, with the aim of quantifying the phenomena and processes resulting from the occurrence of important critical events and their effects. In particular the models consider the effects induced to the health of the exposed subjects, to the infrastructures, and to the surrounding environment, by using thermodynamics and fluid dynamics principles and methods;
- adaptation of available technological devices and systems, characterised in terms of specific reliability and efficiency, to affect the development of the possible accidents trajectories;
- definition of acceptance criteria and risk quantification measures;

© The Author(s) 2019

F. Borghetti et al., *Road Tunnels*, PoliMI SpringerBriefs,
https://doi.org/10.1007/978-3-030-00569-6_1

Fig. 1.1 Interdependence
between the specific models
of calculation

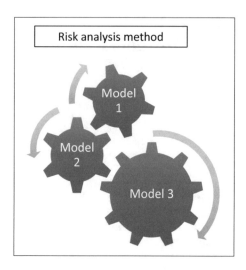

Fig. 1.1 Interdependence between the specific models of calculation

- supply of quantitative information on the risk level of the tunnel system to the managers that have to choose preventive and protective measures to be implemented in order to obtain a risk reduction and control.

Each phase of the method can be developed through specific analytical models that can simulate very complex systems where data needed for studying the problem are often not certain but are influenced by great variability (Fig. 1.1).

In this manner, the behaviour of the system to be studied can be reproduced through mathematical criteria and functions.

The objective of risk analysis is therefore to reproduce or simulate one or more scenarios that could occur in future, following a potential event.

1.1 Risk and Its Components

The concept of risk can be considered complementary to that of safety.

In technical language, there are several and numerous definitions that are commonly used to express the meaning of safety, but all of them confirm that safety must be considered as the *condition of tolerable risk*. The reference term is therefore the safety level that the community reputes as being necessary for reaching a specific life or work situation when carrying out a specific activity [1, 5].

Complete safety is intended as being the limit to which the value of real safety asymptotically tends, even though it can never be reached in any human activity no matter which resources are used [4], as illustrated in Fig. 1.2. This implies that no anthropic activity, including the use of a tunnel, is without risk.

From the concept of safety as a level of tolerable risk, therefore, derives that the risk can be considered as the complementary element to the safety itself; in addition,

Fig. 1.2 Relationship between safety and resources used [4]

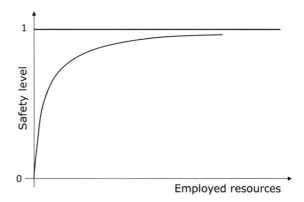

it can be intended as the possibility that a loss of or damage to specific exposed elements can occur following a critical event [6].

In this framework, risk can be expressed according to the probability of occurrence P and the magnitude of the consequences C [7]:

$$R = f(P, C)$$

More in detail, considering the consequences as a combination of vulnerability V and exposure E we obtain:

$$R = f(P, V, E)$$

where:

- the probability of occurrence P over time and in space expresses the frequency at which a critical event of given intensity occurs in a system;
- the vulnerability V represents the propensity of the components of a system to suffer a damage as a consequence of a critical event;
- the exposure E is the quantification of the elements that are subject to risk and which can potentially be damaged.

The combination between vulnerability V and exposure E makes it possible to define the consequences or magnitude of the considered event, namely the quantification of the damage caused by the event according to its intensity.

1.2   Risk Analysis Method

In general, a risk model can be defined through an *application* between the set of dangerous events and the set of consequences [8], as shown in Fig. 1.3.

Fig. 1.3 Representation of
the risk model through the
set theory [8]

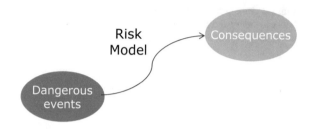

Both these sets are of the probability type, and the set of consequences defines the potential damage that can be associated with a system, that could be affected by dangerous events.

The set of consequences and their magnitude depend on the measures and mitigation actions implemented at both design and operational levels that involve the following factors:

- human behaviour;
- structural solutions;
- technological systems;
- operating and control procedures.

The risk analysis process can, therefore, be divided into phases as shown in Fig. 1.4, and an analytical model can correspond to each phase.

Estimation of the probability of occurrence of an event (frequency) is generally the most critical phase in the risk analysis procedure, and it is also difficult to quantify. On the contrary, the consequence analysis is considered less subject to macroscopic errors.

Tunnel risk analysis is therefore an analytic method that fundamentally consists in identifying the answers to the following main questions [9, 10]:

- what could happen inside the *tunnel system*?
- what is the probability of occurrence of the event?
- having established that the event occurs, what are its possible consequences?

The first question implies the definition of a scenario characterised by at least one hazardous property. The second expresses the probability that a specific scenario occurs, while the last faces consequences quantification.

From this perspective, risk analysis can be seen as a useful instrument for supporting decisions when evaluating the safety of the *tunnel system*, identifying those infrastructure, equipment and management procedures which guarantee greater benefits in terms of expected risk reduction and at the same cost.

Safety in tunnels can be improved through actions concerning infrastructures, equipment and management procedures. The identification and evaluation of these provisions requires an analysis that can make the costs and benefits of each action emerge, so that a choice (decision) can be made on the effectiveness, priority and sequence of actuation of these safety measures. A decision-making process can be

Fig. 1.4 Risk analysis structure

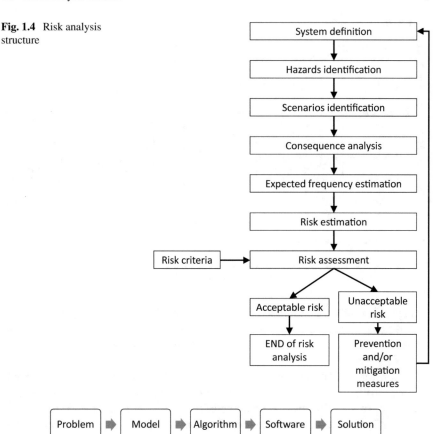

Fig. 1.5 Decision-making process in risk analysis [11]

broken down schematically into five phases as shows in the flow chart presented in Fig. 1.5 [10].

1.3 Risk Representation and Acceptance

When analysing risks in tunnels, the risk is represented through curves on the F-N plane that relates the frequency of dangerous events (F) and the consequences expressed in terms of fatalities (N).

These curves represent the *societal risk*, defined as the number of people who can be affected by a certain damage (in this case death). These curves are determined considering the number of people involved in the accident (event) and the duration of their exposure to the potential damage.

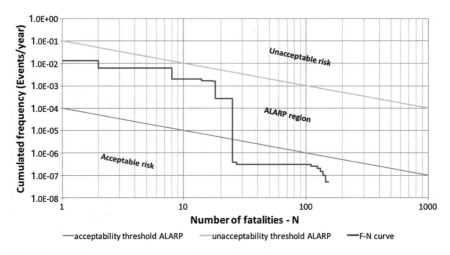

Fig. 1.6 Representation of risk on the F-N plane (blue curve)

Figure 1.6 shows an example of a risk curve (in blue) on the F-N plane.

If we consider death as the concerned damage in a given context/system, the F-N curves supply information on the cumulated frequency of events that can cause at least N fatalities.

To construct F-N curves for each accident/event, identified as reasonably probable to occur inside the tunnel, it is necessary to evaluate the expected frequency of occurrence of the event and the total number of people that are exposed to it.

This information can be used to calculate the cumulated frequency F, namely the frequency at which an accident (any) can be expected to occur and which can cause a number of fatalities equal to or above N, summing the frequency of occurrence of all the accidents able to cause at least N fatalities.

The subsequent step consists in evaluating risk acceptance through the use of recognized risk criteria. In substance, it is necessary to verify if the *risk level* of a specific tunnel system is acceptable or not.

Thus, as previously mentioned, there are no anthropic or natural activities which can be considered absolutely safe or without risk, the prevention and/or the mitigation of risk mean that the risk value can be lowered below certain thresholds of societal acceptance.

Risk acceptance is a very complex matter that has been studied even by sociologists and psychologists, because it involves aspects tied to perception, level of instruction, social status and religion.

Risk in itself cannot be accepted unless compared with the benefit it brings. Considering, for example, the societal risk associated with a road tunnel, mitigation through the introduction of measures (related to infrastructure, equipment and management procedures) must be compared with the possible expected advantages (benefits), for example the reduction in travel time, road accidents, atmospheric and noise pollution.

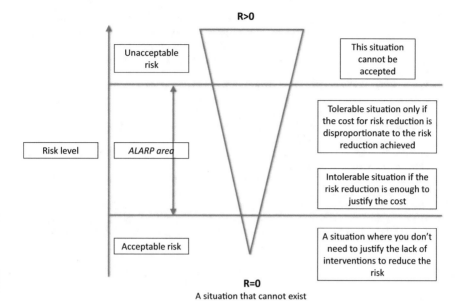

Fig. 1.7 *ALARP* risk acceptance principle

At an international level, there are different consolidated acceptance criteria, among which the *ALARP—As Low As Reasonably Practicable* criterion that involves two thresholds [12], as shown in Fig. 1.7, that determine three risk zones:

- unacceptable risk;
- risk in the *ALARP* zone;
- acceptable risk.

At the basis of this principle is the search for solutions that make it possible to reach the lowest reasonably possible risk level in consideration of the necessary costs.

For a specific tunnel system, the F-N curve must therefore be compared with the curves that represent the risk acceptance criterion and which identify, on the Cartesian plane, also a region considered as unacceptable. Any F-N curve that intersects the unacceptable region represents an unacceptable risk situation for the community as illustrated in Fig. 1.6.

References

1. European Union (2004) Directive 2004/54/EC on Minimum safety requirements for tunnels in the Trans-European Road Network
2. Italian Ministry of Infrastructures and Transports (2006) Adoption of the Directive 2004/54/EC on the safety of tunnels belonging to the Trans-European Road Network. G.U., 235, Oct 2006 (in Italian)

3. Borghetti F, Derudi M, Gandini P, Frassoldati A, Tavelli S (2017) Safety in road tunnels. In: Tunnel fire testing and modeling: the Morgex North tunnel experiment. Springer International Publishing, pp 1–5. https://doi.org/10.1007/978-3-319-49517-0_1
4. Simonetti S (2002) Rischio ambientale—Il rischio della civiltà industriale. Dario Flaccovio Editore (in Italian)
5. Domenichini L, Caputo FJ (2015) Sicurezza in galleria. Egaf edizioni S.r.l. (in Italian)
6. UNDRO Office of the United Nations Disaster Relief Coordinator (1979) Natural disasters and vulnerability analysis, Report of Expert Group Meeting, Geneva
7. International Organizational for Standardization (2014) Safety aspects—guidelines for their inclusion in standards, ISO/IEC GUIDE 51:2014
8. ANAS (2009) Linee Guida per la progettazione della sicurezza nelle Gallerie Stradali secondo la normativa vigente (in Italian)
9. Kaplan S, Garrick BJ (1981) On the quantitative definition of risk. Risk Anal 1:11–27
10. Kaplan S (1997) On the words of risk analysis. Risk Anal 17:407–417
11. Bruglieri M, Colorni C (2012) Ricerca Operativa. Zanichelli Ed. (in Italian)
12. Derudi M, Borghetti F, Favrin S, Frassoldati A (2018) Tram: a new quantitative methodology for tunnel risk analysis. Chem Eng Trans 67:811–816. https://doi.org/10.3303/CET1867136

Chapter 2
Background of Modelling Approaches and Tools

Abstract This chapter illustrates the state of the art of the risk analysis methods and models for road tunnels with particular reference to the users' egress models.

Model design and development were inspired by the *QRAM—Quantitative Risk Assessment Model* software, developed jointly by PIARC[1] and OECD[2] (1997–2001) with the aim to estimate the risk related to the transport of dangerous goods.

In addition, reference was made to scientific researches focused on egress models and CFD (Computational Fluid Dynamics) and zone models for fire dynamics and smoke propagation description.

2.1 QRAM Software

The model compares the risk caused when dangerous goods are transported along alternative roads, through tunnels or on open-air roads. In addition, the effect of some tunnel infrastructure measures can be evaluated: for example, QRAM can be used to compare the effects of a change in the distance between the emergency exits. The model returns individual risk maps along a tunnel and along an open-air road and the F-N curves that represent the societal risk [1–7, 9].

QRAM software is based on Microsoft Excel. There are many parameters that the analyst has to insert to start the program, for example the geometry, the emergency exits, the tunnel equipment and management measures, traffic data, etc.

---

[1]The Permanent International Association of Road Congresses was established on 27 April 1909 as a no-profit association following the first international congress on the road that was held in Paris in 1908, under the patronage of the French government. The association counts 142 member countries, and AIPCR has been present as a consultant in the economic and social committee of the United Nations Organisation (www.piarc.org/fr/) since 1970.

[2]The Organisation for Economic Co-operation and Development is an intergovernmental economic organisation with 35 member countries, founded in 1960 to stimulate economic progress and world trade (www.oecd.org).

© The Author(s) 2019
F. Borghetti et al., *Road Tunnels*, PoliMI SpringerBriefs,
https://doi.org/10.1007/978-3-030-00569-6_2

QRAM considers 13 accidental scenarios involving dangerous goods. It also considers the accidents frequencies (based on historical data), the physical consequences of the accidents, and the effects of heat and smoke.

The software allows users to:

- compare the risks due to dangerous goods transportation along alternative routes; for example, via the tunnel or an alternative open route;
- evaluate issues associated with tunnel regulations; for example, QRAM can be used to support the decision to choose the most relevant of the proposed ADR dangerous goods groups for each specific road tunnel;
- compare the risks along a route with acceptance criteria for individual and societal risks;
- evaluate tunnel equipment options; for example, QRAM can be used to compare the effects of changing the spacing between emergency exits.

The software can be used to perform a risk analysis as required for dangerous goods transportation by the European Directive 2004/54/EC on minimum safety requirements for tunnels in the trans-European road network.

As previously mentioned, QRAM considers 13 accidental scenarios which are representative of key dangerous goods groups:

1. Heavy Goods Vehicle fire with no dangerous goods (20 MW);
2. Heavy Goods Vehicle fire with no dangerous goods (100 MW);
3. Boiling Liquid Expanding Vapour Explosion (BLEVE) of Liquid Petroleum Gas (LPG) in cylinders;
4. Pool fire of motor spirit in bulk;
5. Vapour Cloud Explosion (VCE) of motor spirit in bulk;
6. Release of chlorine in bulk;
7. BLEVE of LPG in bulk;
8. VCE of LPG in bulk;
9. Torch fire of LPG in bulk;
10. Release of ammonia in bulk;
11. Release of acrolein in bulk;
12. Release of acrolein in cylinders;
13. BLEVE of carbon dioxide in bulk (not including toxic effects).

2.2 Theoretical Basis for the Egress Model

The process of evacuation from a tunnel in emergency conditions is a complex phenomenon that involves different factors, tied to both physical characteristics such as the tunnel geometry or the distance between the emergency exits, and human behaviour. While the first types of factor are deterministic and as such easy to insert in a model, the variables tied to human behaviour are difficult to define because of their intrinsic variability. The literature presents studies that analyse the effects that these factors have on the main parameters and processes of the egress model. It was discovered that the egress time is mainly influenced by the user movement speed during evacuation and by its pre-movement time [8]; in addition, the process of choosing the emergency exit by the user is also fundamental. All this is also strongly conditioned by the effects of the accidental event, such as the propagation of toxic gases and reduced visibility due to smoke propagation [9–15].

2.2.1 Pre-movement Time

Pre-movement time is a parameter composed of the time required by each tunnel user to become aware of the danger (recognition time), the time taken to decide on evacuation (response time) and the time taken to exit from the vehicle. It is influenced by human factors tied to the emotional state of the tunnel users, their past experience, and their knowledge of the environment. In addition, it must be considered that users are strongly influenced by the actions of other people and the tunnel conditions, for example visibility caused by smoke or the presence of technical measures such as loudspeakers or variable message panels [11, 12].

The pre-movement time can be calculated in a deterministic manner or using a distribution law. The estimation of the average value of this parameter is still a point of controversy, however. The experiments carried out in the Benelux tunnel showed how the intervention of alarms and verbal messages caused users to exit from their vehicles in around 35 s. The ANAS guidelines indicates a pre-movement time that can vary from 90 to 300 s. These values remain representative, because the lack of reliable data leaves the choice of the pre-movement time value to the analyst, who must consider the above-mentioned aspects [16].

2.2.2 Choice of the Emergency Exit

This process depends on the tunnel conditions, such as the presence of smoke or the distance between the emergency exits. In general, the users tend to evacuate towards the closest exit, excluding the case in which it is not very visible because of smoke presence caused by a fire [16].

The choice of the exit is fundamental because the distance travelled by the user during evacuation and, as a result, the egress time, depends on it. The majority of existing models always considers the closest exit; it is possible, however, to consider the effect that a reduction in visibility caused by the spreading of smoke has on the user's choice of exit [10, 11, 17].

2.2.3 Movement Speed

This parameter is strongly influenced by the visibility conditions inside the tunnel and by the interaction that the users have during the evacuation process.

The movement speed can vary from 1.25 m/s in the case of good visibility, to 0.3 m/s in cases of reduced visibility and in the presence of harmful gases. Today, the use of lower movement speeds is the best way to consider the influence of an accident (and its consequences) on the egress process [13]. In addition, the presence of equipment and measures such as emergency ventilation can have a positive effect on movement speed, slowing the smoke propagation within the tunnel [15].

2.3 Existing Egress Models

There is a large number of software and models dedicated to simulating the egress process of people in emergency conditions from different environments. Developed initially to study the phenomenon of evacuation from civil buildings, they were extended over time to the confined and underground environment represented by tunnels. Their application to tunnels was not immediate because unique environments with their own specific characteristics [18]: underground spaces, unknown to the users, unnatural light, and many other factors that condition different aspects of human behaviour. As a result, some new instruments that were recently presented to the market, among which *EvacTunnel*, *GridFlow*, *STEPS*, *Pathfinder*, *FDS + Evac*, were designed specifically to analyse the safety conditions of tunnel users.

The aim of existing egress models is to estimate the number of fatalities in a tunnel under a defined event scenario. To evaluate the consequences of an accident (no matter whether a fire, an explosion or something else,...) in terms of human life, each model must first quantify the time needed for the single user to move to a safe place (RSET-Required Safe Egress Time), with the aim of comparing it with the maximum time available before a human being reaches the thresholds of toxic substances/lack of oxygen/high temperature prescribed in the acceptance criteria (ASET-Available Safe Egress Time). All those whose RSET < ASET are held to be safe and in a safe place while waiting for the emergency vehicles, while those whose RSET > ASET cannot complete the evacuation process and are considered as victims by the model. Figure 2.1 show the times which characterize the exit from the tunnel of its users.

Three main model categories are defined: behavioural models, movement models and semi-behavioural models. The first includes the decision-making processes of the users and their reactions to the tunnel conditions. The second simulates the movement of the tunnel occupants from one point to another. Finally, the last, by simulating tunnel user movement, implicitly reproduces human behaviour considering a distribution of the pre-movement times, the influence of smoke, etc.

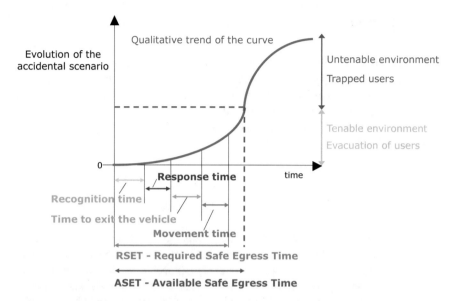

Qualitative trend of the curve

Evolution of the
accidental scenario

Untenable environment

Trapped users

Tenable environment

Evacuation of users

0

Response time

time

Recognition time

Time to exit the vehicle

Movement time

RSET - Required Safe Egress Time

ASET - Available Safe Egress Time

Fig. 2.1 Representation of characteristic times in the evacuation process of users from the tunnel

2.3.1 EvacTunnel

EvacTunnel simulates the evacuation process for one-way tunnels during an emergency (accidents, fire, terrorist attack, etc....). For two-way tunnels, the software can be used twice, once for each direction. The software is based on a stochastic simulation process, through the Monte Carlo method: different simulations can be carried out in a few seconds, obtaining a statistical distribution of the output value. *EvacTunnel* considers the effects of smoke inside the tunnel and yields the number of dead and injured users. The model also gives the total number of trapped vehicles and the user egress time, supplying statistical values such as the average and the variance [12, 18].

2.3.2 GridFlow

This model was implemented mainly for buildings but can also be used for tunnels. The model defines a continuous spatial system that uses a grid to represent the distance to the emergency exit. *GridFlow* manages to simulate each separate user with his relative parameters, for example the pre-movement time and the movement speed, using a statistical distribution function [13, 14, 16].

2.3.3   STEPS

This software can simulate the movement of pedestrians and their behaviour in relation to the environment they find themselves in, under normal and emergency conditions. It uses a 0.5×0.5 m grid where each user occupies a cell; user movement
depends on the availability of the next cell. The model parameters, such as the pre-
movement time and the movement speed, are implemented through a distribution
function [13–16].

2.3.4   PathFinder

Pathfinder simulates the evacuation process in emergency conditions. It includes an
integrated interface and presents animated 3D results. This software can estimate all
the evacuation parameters and supplies them in graphical form. The analytical model
can be coupled with a fire simulation model in order to carry out a risk analysis. Even
in this case the parameters are represented through a distribution function [13–16].

2.3.5   FDS + Evac

FDS + Evac is a semi-behavioural model that combines an egress simulation model
with a fluid dynamic model where the fire and the occupants being evacuated interact.
FDS + Evac considers each occupant in the evacuation area as a separate agent,
using stochastic properties to assign characteristics to them. The model also supplies
movement speed values and the doses of toxic gas inhaled in the computational
domain [13–16, 19–21].

Figure 2.2 shows an example of smoke propagation in a tunnel simulated using
FDS code [22].

The described egress models are potentially very effective for studying risk analysis in tunnels, in spite of the fact that the factors tied to human behaviour can cause
difficulty in calibrating the input parameters. The analysis of behavioural uncertainty
can be used to consider the impact of the unexplained variance in human behaviour
on the simulation results produced by FDS + Evac [23].

If the analyst does not have enough experience with assumptions made from the
model, the risk of inaccurate and uncertain analyses remains [13].

The egress model that is implemented and described in this document is *deterministic* and *macroscopic*. It is deterministic because the variables used are considered
as being known and as such they cannot be attributed with a variance related to reference values or a distribution of probability, carrying out a "sure" study on system
operation and development without the use of stochastic variables. Macroscopic,
instead, refers to the aggregation level of the variables used [21].

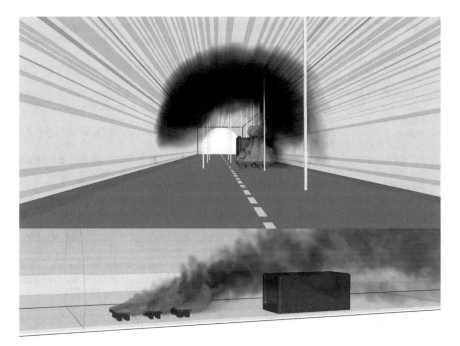

Fig. 2.2 Example of smoke propagation simulation using FDS

References

1. OECD (2001) Safety in tunnels—transport of dangerous goods through road tunnels. Organization for Economic Co-operation and Development—OECD Publications, Paris
2. Saccomanno F, Haastrup P (2002) Influence of safety measures on the risks of transporting dangerous goods through road tunnels. Risk Anal 22:1059–1069
3. Lacroix D, Cassini P, Hall R, Saccomanno F (1999) Transport of dangerous goods through road tunnels: an integrated QRA Model developed under the joint OECD/PIARC Project ERS2. International ESReDA Seminar on 'Safety and Reliability in Transport', Oslo
4. Ruffin E, Cassini P, Knoflacher H (2005) Transport of hazardous goods (Chap. 17). Beard A, Carvel R (eds) The handbook of tunnel fire safety. Thomas Telford Ltd., London
5. Knoflacher H, Pfaffenbichler PC (2004) A comparative risk analysis for selected Austrian tunnels. In: Proceedings of 2nd international conference tunnel safety and ventilation, Graz
6. Knoflacher H, Pfaffenbichler PC, Nussbaumer H (2002) Quantitative risk assessment of heavy goods vehicle transport through tunnels—the tauerntunnel case study. In: Proceedings of 1st international conference tunnel safety and ventilation, Graz
7. Knoflacher H, Pfaffenbichler P (2001) A quantitative risk assessment model for road transport of dangerous goods. In: Proceedings of the 80th annual meeting of the transportation research board, Washington DC
8. Lovreglio R, Ronchi E, Nilsson D (2015) A model of the decision-making process during pre-evacuation. Fire Saf J 78:168–179
9. Derudi M, Borghetti F, Favrin S, Frassoldati A (2018) TRAM: a new quantitative methodology for tunnel risk analysis. Chem Eng Trans 67:811–816. https://doi.org/10.3303/CET1867136
10. Caliendo C, De Guglielmo ML (2016) Quantitative risk analysis based on the impact of traffic flow in a road tunnel. Int J Math Comput Simul 10:39–45

11. Caliendo C, Ciambelli P, De Guglielmo ML, Meo MG, Russo P (2012) Simulation of people evacuation in the event of a road tunnel fire. SIIV—5th Int Congr—Sustain Road Infrastruct Soc Behav Sci 53:178–188
12. Capote JA, Alvear D, Abreu O, Cuesta A, Alonso V (2012) A real-time stochastic evacuation model for road tunnels. Saf Sci 52:73–80
13. Ronchi E, Colonna P, Berloco N (2013) Reviewing Italian fire safety codes for the analysis of road tunnel evacuations: advantages and limitations of using evacuation models. Saf Sci 52:28–36
14. Ronchi E (2013) Testing the predictive capabilities of evacuation models for tunnel fire safety analysis. Saf Sci 59:141–153
15. Ronchi E, Alvear D, Berloco N, Capote J, Colonna P, Cuesta A (2010) Human behavior in road tunnel fires: comparison between egress models (FDS + Evac, Steps, Pathfinder). In: International conference on fire science and engineering Interflam, pp 837–848
16. Ronchi E, Colonna P, Capote J, Alvear D, Berloco N, Cuesta A (2012) The evaluation of different evacuation models for assessing road tunnel safety analysis. Tunn Undergr Space Technol 30:74–84
17. Lovreglio R, Fonzone A, Dell'Olio L (2016) A mixed logit model for predicting exit choice during building evacuations. Transp Res Part A: Policy Pract 92:59–75
18. Alonso V, Abreu O, Cuesta A, Alvear D (2014) An evacuation model for risk analysis in Spanish road tunnels. In: XVIII Congreso Panamericano de Ingeniería de Tránsito, Transporte y Logística (PANAM 2014), Social and Behavioral Sciences, vol 162, pp 208–217
19. Bosco D, Lovreglio R, Frassoldati A, Derudi M, Borghetti F (2018) Queue formation and evacuation modelling in road tunnels during fires. Chem Eng Trans 67:805–810. https://doi.org/10.3303/CET1867135
20. Glasa J, Valasek L (2014) Study on applicability of FDS + Evac for evacuation modeling in case of road tunnel fire. Res J Appl Sci Eng Technol 7:3603–3615
21. Seike M, Kawabata N, Hasegawa M (2017) Quantitative assessment method for road tunnel fire safety: development of an evacuation simulation method using CFD-derived smoke behavior. Saf Sci 94:116–127
22. Borghetti F, Derudi M, Gandini P, Frassoldati A, Tavelli S (2017) Evaluation of the consequences on the users safety. In: Tunnel fire testing and modeling: the Morgex North tunnel experiment. Springer International Publishing, pp. 65–75. https://doi.org/10.1007/978-3-319-49517-0_5
23. Lovreglio R, Ronchi R, Borri D (2014) The validation of evacuation simulation models through the analysis of behavioural uncertainty. Reliab Eng Syst Saf 131:166–174

Chapter 3
Model Structure

Abstract This chapter describes the logical structure of the proposed tunnel risk analysis model. In particular, the event tree technique for estimating the frequencies of the accidental scenarios and the procedure for defining the location of the accidental scenarios along the tunnel are illustrated.

As already discussed in previous chapter, risk analysis is a methodology aimed at evaluating and managing the risk associated with a specific tunnel system compared to the consequences on the exposed population. Risk evaluation is a process that identifies the sources of danger and determines the exposure of the population to their possible effects, including also an estimation of the connected uncertainties.

The quantitative risk analysis of a road tunnel consists of implementing a model that can calculate a complementary cumulated curve on the F-N Frequency-Fatality plane, where each point represents the frequency of occurrence of an accidental scenario to which the number of user's fatalities is associated. Risk acceptance is obtained using the *ALARP—As Low As Reasonably Practicable* criterion that identifies the portion of the F-N diagram included between the risk acceptance limit and its tolerability level, within which the cost-benefits analysis is applied as a guiding criterion when making risk management decisions.

Risk management involves the decision-making process, following risk evaluation, related to the definition and realisation of safety measures in a manner that is consistent with the characteristics of the social, economic and political context of the country in which the structure is made. Figure 3.1 shows the logical outline of the implemented risk analysis model.

The following chapters illustrate the implemented risk analysis model in detail.

© The Author(s) 2019
F. Borghetti et al., *Road Tunnels*, PoliMI SpringerBriefs,
https://doi.org/10.1007/978-3-030-00569-6_3

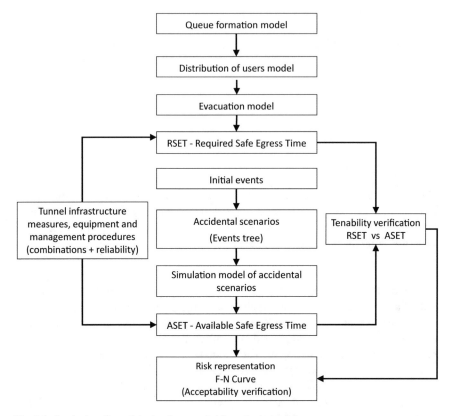

Fig. 3.1 Logical outline of the implemented risk analysis model

3.1   Event Tree Analysis and Frequencies of Occurrence of Accidental Scenarios

In the risk analysis process for tunnels, the use of the *Bow-tie diagram*, recognised in the literature on industrial process safety, makes it possible to identify the fault tree and the event tree separated by the starting event [1–5].

The fault tree identifies and represents the possible causes of an initial event, while the event tree represents their development, as shown in Fig. 3.2.

The prevention measures aimed at reducing the probability of the occurrence of an initial event are therefore positioned on the left side of the diagram, while the protection measures, the purpose of which is to mitigate the possible consequences, are on the right side.

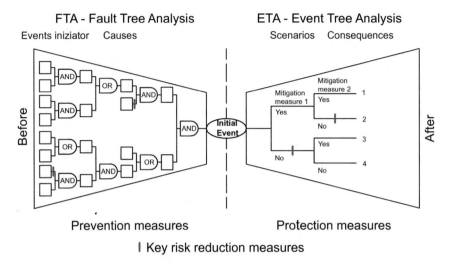

Fig. 3.2 Representation of the *Bow-tie* approach used for safety evaluations in the process [1]

The model takes into consideration two initial events, fire and dangerous material spills, characterised by an initial occurrence frequency, *Ffire* and *Frel*. These events are typical and representative of the potential dangers inside the tunnel system as a confined environment. Indeed, in agreement with what is indicated in the European Directive 2004/54/EC and the Italian Legislative Decree 264/2006, the events of road accidents connected with the geometric characteristics of the infrastructure and not induced by the specific tunnel environment (confined), should not be considered in the tunnel risk analysis. In fact, this kind of events does not cause risks other than those already connected with road circulation, and thus are to be considered only for prevention in the traffic regulation and road design context. For this reason, the victims of this type of accident are recorded as caused by ordinary road accidents.

At a general level, some possible primary causes of the occurrence of these initial events follow:

- faults in the mechanical parts of the motor or, more in general, the vehicle, for example brake system overheating, turbocharger explosion, gears;
- collision between vehicles, for example because the safety distance was not kept, or between an unmoving, broken down vehicle and moving traffic;
- collision between vehicles and tunnel elements/devices (pavements, walls, guard rails, etc.);
- collision following the overturning of a vehicle caused, for example, by a tyre puncture or tyre explosion, high speed, etc.;
- loss of fuel from the vehicle fuel tank;
- container breakage (spills of flammable, toxic, corrosive, etc. material).

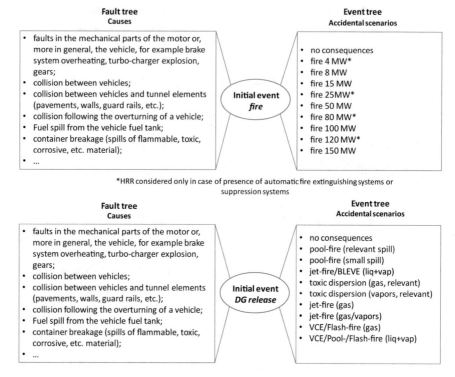

Fig. 3.3 Bow-tie diagram of the event: fire starting (above) and DG release (below)

Figure 3.3 shows the bow-tie diagram of the initial event: fire and DG (Dangerous Goods) release.

Through the use of the event tree, made up of a sequence of questions with Boolean answer, it is possible to estimate the development of the two initial events in terms of accidental scenarios—SC. The two initial frequencies, *Ffire* and *Frel*, evaluated on the basis of the available and observed traffic data, are multiplied by probabilities associated with each question.

This technique can be used to determine the frequencies of occurrence F_s associated to each scenario SC_s, for all the possible total S accidental scenarios.

Figure 3.4 shows an event tree with an example for calculating F_s starting from an initial frequency F_{in} related to a generic event.

The frequency of occurrence of an accidental scenario, say F_8, is determined as follows:

$$F_8 = F_{in} * Pn_1 * Pn_2 * Pn_3 * Pn_4 * Py_5 * Py_6$$

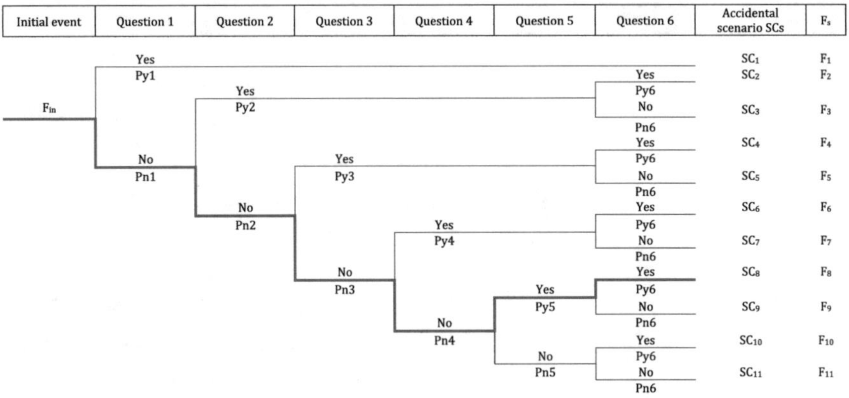

Fig. 3.4 Example of an event tree structure for accidents. *Py* and *Pn* represent the probability of yes/no

Figure 3.5 shows a numeric example of a risk tree which refers to the initial DG release, from which the different scenarios SC_s can be obtained with associated frequency F_s starting from the initial frequency of the DG release *Frel*.

Table 3.1 shows the estimated HRR associated with the different categories of vehicle.

Figure 3.6 presents the characteristics of some typical fires associated with light vehicles (cars), buses, heavy vehicles and vehicles used to transport dangerous goods (e.g. oil tankers or trailers). An indication of the maximum fire intensity that can be reached is given for each type of fire, in terms of heat release rate (HRR) and the corresponding estimated smoke flowrate [1, 3–7].

Figure 3.7 illustrates an example of an event tree that gives the different scenarios SC with associated frequency F_s, starting from the initial frequency of the fire event *Ffire*.

Using the event tree, the model is currently able to analyse 14 accidental scenarios, 5 of which are associated with fire and 9 with DG release as shown in Table 3.2. Additional fire HRR (4, 25, 80 and 100 MW) are considered in the case of mitigation measures such as the automatic fire extinguishing system or suppression systems that can reduce the fire intensity. In this case the total number is 18.

Each *s-th* scenario develops, therefore it evolves inside the tunnel more or less quickly according to the fire intensity or the quantity of spilled hazardous material.

To know how each accidental scenario will develop, Tf_s is defined as the progress time of the effects of the *s-th* scenario: it is fundamental for calculating fatalities because it is compared with the user survival times, as described in the following.

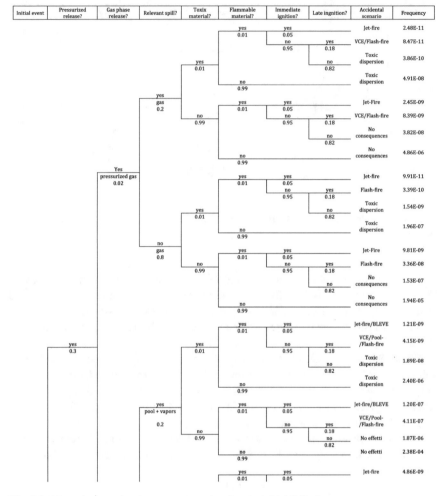

Fig. 3.5 Numeric example of an event tree related to the initial DG release

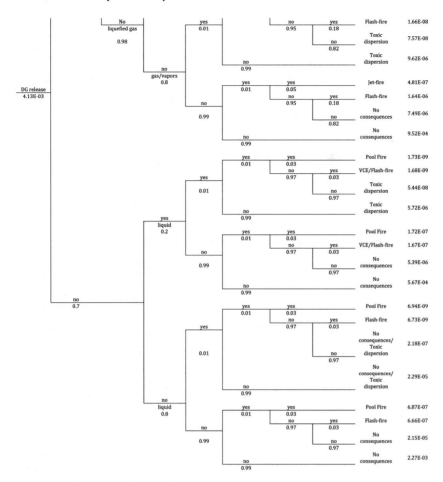

Fig. 3.5 (continued)

Table 3.1 Types of vehicle considered and respective HRR

Type of vehicle	HRR (MW)
Light vehicle—LV1	8
Light vehicle—LV2	15
Heavy vehicle—HV1	50
Heavy vehicle—HV2	100
Heavy vehicle—DG	150

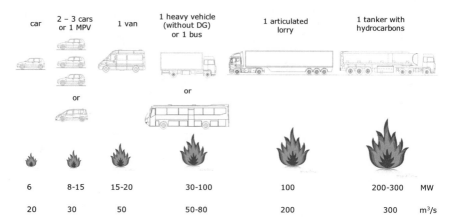

Fig. 3.6 Fire HRR and smoke production for different types of vehicle [1]

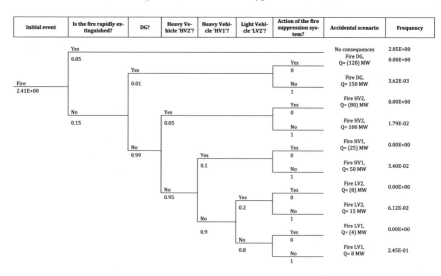

Fig. 3.7 Numeric example of an event tree related to the initial fire event [2]

3.2 Position of the Accidental Scenarios

The calculation model predicts that the *s-th* scenario can occur at different points in the tunnel: in this manner, it is assumed that the source of the initial event can be located in a generic position PO_p as a percentage of the total length of the tunnel *Ltot*.

As a result, *O* positions of PO_p can be analysed for each accidental scenario SC_s. As shown in Fig. 3.8, the model can analyse up to 18 possible positions (and it uses a default value $O = 6$).

Table 3.2 Accidental scenarios and numerical example of frequencies of occurrence associated with the initial fire and DG release events (in this example the automatic fire extinguishing systems or suppression systems are not present)

Initial event	N. accidental scenario	Accidental scenario SCs		Frequency of occurrence Fs
Fire	1	F1	Fire LV1, Q = 8 MW	2.45E−02
	2	F2	Fire LV2, Q = 15 MW	6.12E−03
	3	F3	Fire HV1, Q = 50 MW	3.40E−03
	4	F4	Fire HV2, Q = 100 MW	1.79E−03
	5	F5	Fire DG, Q = 150 MW	3.62E−04
DG release	6	R1	Pool-fire (relevant spill)	1.42E−07
	7	R2	Pool-fire (small spill)	5.70E−07
	8	R3	Jet-fire/BLEVE (liq + vap)	9.97E−08
	9	R4	Toxic dispersion (gas, relevant)	4.06E−08
	10	R5	Toxic dispersion (vapors, relevant)	1.99E−06
	11	R6	Jet-fire (gas)	2.03E−09
	12	R7	Jet-fire (gas/vapors)	4.07E−07
	13	R8	VCE/Flash-fire (gas)	6.96E−09
	14	R9	VCE/Pool-/Flash-fire (liq + vap)	3.41E−07
Fire*	15	F6	Fire LV1, Q = 4 MW	0.00E+00
	16	F7	Fire HV1, Q = 25 MW	0.00E+00
	17	F8	Fire HV2, Q = 80 MW	0.00E+00
	18	F9	Fire DG, Q = 120 MW	0.00E+00

*Accidental scenarios and HRR considered only in case of presence of automatic fire extinguishing systems or suppression systems

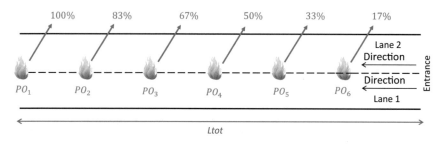

Fig. 3.8 Examples of positions PO_p at which the accidental scenario can occur

3.2.1   Probability of Occurrence of Accidents for the Different Positions Along the Tunnel

The proposed model makes it possible to associate the $Ppos_p$ parameter, which represents the probability of occurrence of the $s\text{-}th$ accidental scenario at the $p\text{-}th$ position, with each position PO_p. The model considers up to 18 different positions, each characterized by a probability of occurrence defined by the user.

References

1. Borghetti F, Derudi M, Gandini P, Frassoldati A, Tavelli S (2017) Safety in road tunnels, in Tunnel fire testing and modeling: the Morgex North tunnel experiment. Springer International Publishing, pp 1–5. https://doi.org/10.1007/978-3-319-49517-0_1
2. Derudi M, Borghetti F, Favrin S, Frassoldati A (2018) TRAM: a new quantitative methodology for tunnel risk analysis. Chem Eng Trans 67:811–816. https://doi.org/10.3303/CET1867136
3. Beard A, Carvel R (2005) The handbook of tunnel fire safety. Thomas Telford Publishing
4. PIARC (1999) Fire and smoke control in road tunnel. Permanent International Association of Road Congress
5. NFPA 502 (2004) Standards for road tunnels, bridges and other limited access highways, 2004 Ed., National Fire Protection Association, Quincy, Massachusetts
6. BD 78/99—Design manual for roads and bridges (1999) The Highways Agency
7. Maevsky IY (2011) Design fires in road tunnels. A Synthesis of Highway Practice, NCHRP SYNTHESIS 415

Chapter 4
Tunnel Infrastructure Measures, Equipment and Management Procedures

Abstract This chapter describes the interdependence and estimate the reliability of the tunnel infrastructure measures, equipment and management procedures considered in the proposed model.

The aim is to define the total number of tunnel infrastructure measures, equipment and management procedures MI_m that can be analysed using the model.

Each measure influences a specific model parameter. For example, the presence of a mechanical ventilation system and emergency lighting influence the user egress speed.

The measures can also influence the frequencies of occurrence F_s of the *s-th* accidental scenarios: the presence or lack of the *m-th* measure can, in fact, modify the probabilities associated with the questions in the event tree.

. Two examples are given for the following measures considered in the model:

- flammable liquid drainage;
- water supply system;
- specialised rescue and fire fighting team.

In the first example, illustrated in Figs. 4.1 and 4.2, the drainage of flammable liquids influences the event tree related to dangerous goods spills at the question *"Relevant spill?"*: according to whether this measure is available or not, the probability associated with the question varies and, as a result, the frequencies of occurrence F_s of the scenarios associated with dangerous material spills change.

In the second example in Fig. 4.3, it can be seen how the water supply system and the specialised rescue team influence the event tree regarding fires at the question *"Is the fire rapidly extinguished?"*. The respective probability values can be associated with four different cases that are reported further on.

For a generic tunnel, My defines the number of infrastructure measures, equipment and management procedures present from among the total M considered by the model.

In addition, a reliability parameter RM_m is introduced for each *m-th* measure; first of all this parameter depends on the presence of the measure and then on its availability when it is needed.

© The Author(s) 2019
F. Borghetti et al., *Road Tunnels*, PoliMI SpringerBriefs,
https://doi.org/10.1007/978-3-030-00569-6_4

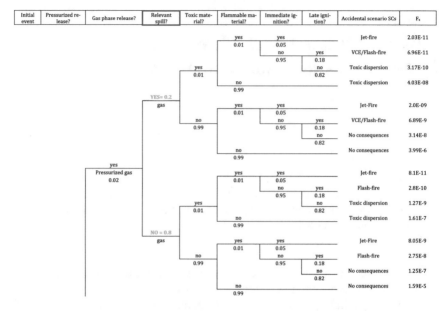

Fig. 4.1 Extract of the event tree: presence of the drainage of flammable liquids

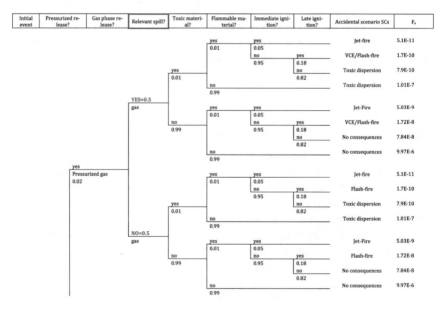

Fig. 4.2 Extract of the event tree: no drainage of flammable liquids

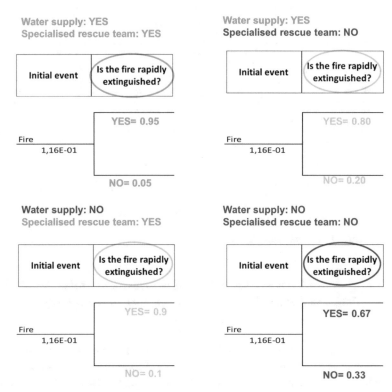

Fig. 4.3 Extract of the event tree. Influence of the fire extinguishing system and the specialised rescue team on the event tree related to fires

In this manner, not only the presence of a specific measure, but also its availability or non-availability because of one or more faults for example, can be considered.

If the generic measure MI_m is present, RM_m can take on the following values:

$$RM_m = \begin{cases} Pry, & if \ MI_m \ is \ available \\ Prn, & if \ MI_m \ is \ not \ available \end{cases}$$

The reliability parameter RM_m can be used to express the probability that the m-th measure is available or not, for example because of a direct or indirect fault (e.g. lack of electricity).

In the event that MI_m is not present, RM_m is always equal to 0.

As RM_m is a probability, $Pr_y + Pr_n = 1$.

A numeric example of RM_m for a specific system measure follows:

$$MI_m = Emergency \ mechanical \ ventilation$$

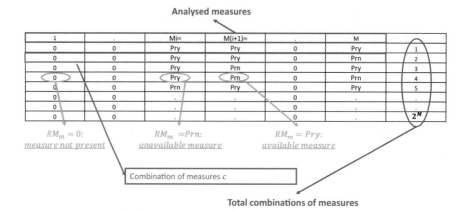

Fig. 4.4 Matrix of the combination of measures c_m

$$RM_{vent} = \begin{cases} Pry = 0.95, & \textit{if } MI_{vent} \textit{ is present and available} \\ Prn = 0.05, & \textit{if } MI_{vent} \textit{ is present but not available} \\ 0, & \textit{if } MI_{vent} \textit{ is not present} \end{cases}$$

In this case, if the emergency mechanical ventilation was present, it would have a 95% probability of being available ($RM_{vent} = 0.95$) in the case of an accidental event, while it would have a 5% probability of not being available ($RM_{vent} = 0.05$); if it was not present, $RM_{vent} = 0$.

After having defined the reliability parameter of each measure RM_m, a matrix can be created with all the possible combinations of RM_m according to the presence and availability of the generic system measure MI_m.

Each line in the matrix represents a combination c_m of RM_m that can be analysed for each *p-th* position of each *s-th* scenario, as shown in Fig. 4.4.

Figure 4.5 gives a numeric example of the previously described matrix that considers 6 measures MI_m for a specific tunnel:

- control centre;
- event automatic identification + camera;
- event automatic identification;
- traffic lights at the tunnel entrance;
- GSM coverage;
- emergency stations.

Of the 6 measures considered, only 3 are present (automatic event identification, traffic lights at the tunnel mouths and emergency stations): the possible combinations c_m can therefore be deduced as 2^3.

Each line of the matrix therefore represents a combination c_m of measures from the 2^{My} that the model can analyse. The columns, instead, represent the reliability parameter RM_m that the individual measure assumes with each $c_m - th$ combination.

Analysed measures

Control centre	Automatic fire detection + video monitoring system	Automatic fire detection	Traffic signals at the entrances	GSM	Emergency stations	H_{cm}
NO	NO	YES	YES	NO	YES	
1	1	0,95	0,95	1	0,95	8,57E-01
1	1	0,95	0,95	1	0,05	4,51E-02
1	1	0,95	0,05	1	0,95	4,51E-02
1	1	0,95	0,05	1	0,05	2,38E-03
1	1	0,05	0,95	1	0,95	4,51E-02
1	1	0,05	0,95	1	0,05	2,38E-03
1	1	0,05	0,05	1	0,95	2,38E-03
1	1	0,05	0,05	1	0,05	1,25E-04

$RM_m = 0$:
measure not present

$RM_m = Prn = 0,05$:
unavailable measure

$RM_m = Pry = 0,95$:
available measure

Combination of measures c

Fig. 4.5 Matrix with different combinations of measures ic_m: values of RM_m and H_{cm}

For a specific position PO_p at which the scenario SC_s can occur, it is therefore possible to analyse 2^{My} combinations of $cm - th$ infrastructure measures, equipment and management procedures.

In addition, as shown in Fig. 4.5, for each $cm - th$ measure combination it is possible to define the probability H_{cm}:

$$H_{cm} = \prod_{cm=1}^{My} RM_m$$

where:

- H_{cm} is given by the product of RM_m of each m-th measure and represents the probability of each combination according to the reliability;
- RM_m is the reliability parameter of each m-th measure that is present.

The sum of the value of H_{cm} relative to each $cm - th$ combination is equal to:

$$\sum_{cm=1}^{2^{My}} H_{cm} = 1$$

The combinations of measures that have a very low probability H_{cm} can be neglected because their contribution does not influence the analysis greatly. Therefore, it is reasonable to define a cumulated probability threshold H_{cm} *cumulated max* < 1: the combinations of analysed measures are such that:

$$\sum_{cm=1}^{cm\,tot} H_{cm} = H_{cm}\ cumulated\ max$$

where *cm tot* is the number of measure combinations analysed and is lower that 2^{My}. H_{cm} *cumulated max* can, for example, be equal to 0.95 or 0.97.

The number of fatalities is estimated for each accidental scenario SC_s that can occur at the generic positions PO_p considering the $c_m - th$ measure combination using the following models:

- queue formation model; it can be used to estimate the queue length of the vehicles that have stopped in the tunnel following an accidental scenario SC_s for each *i-th* lane according to the system and infrastructure measures; MI_m;
- user distribution model; it allows to discretize in cells *j* the queue of vehicles that have stopped in each *i-th* lane and to estimate the number of users in each cell according to the composition of the traffic (e.g. light vehicles, heavy vehicles and buses) and the relative occupation coefficient;
- tunnel user egress model; it simulates the evacuation process of the users who are inside the tunnel following a specific accidental scenario. SC_s. The model can estimate the number of fatalities by comparing the maximum permanence time of the users in the tunnel (*ASET*—Available Safe Egress Time) with the time needed for each occupant to escape and reach a safe place *(RSET*—Required Safe Egress Time). *ASET* considers the characteristic time of propagation Tf_s of the scenario effects, while *RSET* represents the time needed to reach the emergency exits.

The number of fatalities N therefore results as being dependent on the accidental scenario SC_s, on the position PO_p in which the scenario occurs, and on the presence of the infrastructure measures, equipment and management procedures MI_m that determine the $cm - th$ combination.

$$N = N_{m,p,s}$$

Figure 4.6 shows an example of the logic strategy used to calculate N in which two accidental scenarios SC_s, two positions PO_p and four measure combinations c_m $(MI_m = 2)$ are considered: 16 values of $N_{m,p,s}$ can be observed.

4.1 Interdependence Between the Measures

The system measures MI_m are not all independent because the operation of some of them depends on that of the others. On this matter, the following measure classes can be identified:

- Primary measures: this is the first measure class on which the others depend. Without a primary measure, the others could not operate. An example of a primary measure could be the automatic fire identification system, or the emergency stations inside the tunnel;
- Intermediate measures: these allow the primary measures to implement the secondary measures (which will be explained further on) and only work if at least one

Fig. 4.6 Logic strategy for estimating $N_{s,p,c}$. The dependencies between SC_s, PO_p and cm are illustrated

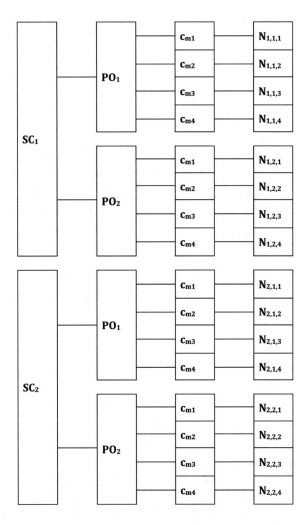

of the primary measures is operational; an example could be the control centre, which would not be operational if there was no primary measure. Two intermediate measures are considered in the model, with the presence of one excluding that of the other. The operation of every intermediate measure is influenced by all the primary measures; as a result, if only one part of the primary measures is present, the generic intermediate measure is not very effective;

• Secondary measures: their operation is tied to that of the intermediate and primary measures. Each secondary measure is influenced by the primary measures and by the intermediary ones, which are the communication bridge between the two classes.

The most critical case is represented by the total absence of all the primary measures; this would imply the inefficiency of all the secondary measures. As an example,

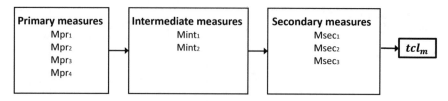

Fig. 4.7 Example of dependence between the primary, intermediate and secondary measures on the calculation of tcl_m

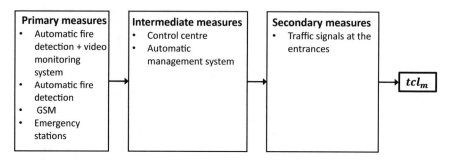

Fig. 4.8 Measures that act on the closing time tcl_m

external traffic lights need a primary measure to activate, otherwise they could not operate.

The outline shown in Fig. 4.7 represents an example that shows the dependence between the classes of measures that influence the closing time of a tunnel tch_m: they represent the time interval that occurs between the occurrence of the accidental event and the activation of the device (e.g. traffic light or VMP) located at the tunnel entrance that indicates its closure.

Each intermediate measure depends on the primary measures and in turn each secondary measure depends on the intermediate measures. In this manner, we can represent the influence that the measures have on the closing time tcl_m.

Figure 4.8 shows an example with real measures. It can be seen that tcl_m is influenced directly by the traffic lights outside the tunnel, the operation of which depends on the primary and secondary measures: without any of them, the external traffic lights could not be activated.

If, instead, these measures were present only in part, the external traffic lights would take a longer time to carry out their task; as a result, tcl_m would have a higher value.

The first step consists of defining a base value of the tunnel closing time $tclbase$. This value represents the minimum time for closing the tunnel following identification of an initial event (e.g. fire or DG release).

Subsequently, a multiplier coefficient $Csec_m$ is defined for each secondary measure, variable between 1 and a maximum of $Csecmax_m$, and adopted to calculate the closing time tch_m:

$$tcl_m = tclbase * \prod_{m=1}^{Msec} Csec_m$$

$Csecmax_m$ can be used to estimate the time required to stop vehicle access to the tunnel independently from the measures that are present.

Then different coefficients have to be defined, $Cint_m$, for each intermediate measure and Cpr_m, for each primary measure, respectively.

The following 2 cases can be distinguished:

(1) if the generic secondary measure is not present $Csec_m = Csecmax_m$.

In the example shown in Fig. 4.8, the external traffic lights are the only secondary measure that influences the closing time; if they were not present the result would be:

$$tcl_m = tclbase * Csecmax_m$$

This is a numeric example of what has just been explained:

$$tclbase = 180\,s; \quad Csec_m = Csecmax_m = 2,5$$

$$tcl_m = 180\,s * 2,5 = 450\,s$$

(2) if, instead, the secondary measure is present, the intermediate and primary measures must be considered.

As already anticipated, a coefficient Cpr_m that can assume the following values is defined for each primary measure:

$$Cpr_m = \begin{cases} Cprpres_m, & \text{if the primary measure is present} \\ 0, & \text{if the primary measure is absent} \end{cases}$$

The sum of the values of all the $Cprpres_m$ is equal to 1:

$$\sum_{m=1}^{Mpr} Cprpres_m = 1$$

• where Mpr represents the total number of primary measures among the generic measures MI_m.

Regarding the primary measures in Fig. 4.8, the following values can be assigned to Cpr_m, as shown in Fig. 4.9.

In this example, the presence of automatic event identification with cameras excludes the presence of that without cameras and vice versa. The first is more efficient therefore the sum of the $Cprpres_m$ of automatic event identification with

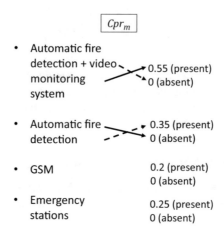

Cpr_m

- Automatic fire detection + video monitoring system → 0.55 (present) / 0 (absent)

- Automatic fire detection → 0.35 (present) / 0 (absent)

- GSM → 0.2 (present) / 0 (absent)

- Emergency stations → 0.25 (present) / 0 (absent)

Fig. 4.9 Numerical example of values of the parameter Cpr_m regarding the primary measures (continuous and dotted arrows indicate mutually exclusive primary measures)

$Cint_m$ → Control centre = 1 / Automatic management system = 0.6

Fig. 4.10 Numerical example of values of the $Cint_m$ parameter regarding intermediate measures. The presence of the control centre excludes the automatic management system and vice versa

cameras, GSM coverage and emergency stations is equal to 1, while if automatic event identification without cameras was present, the sum would be equal to 0.8.

Similarly, to what was done for the primary and secondary measures, each intermediate measure is associated with a coefficient $Cint_m$ variable between 0 and 1. With reference to Fig. 4.8 we can, for example, hypothesise the following values of $Cint_m$ given in Fig. 4.10.

After having defined the coefficients of the primary and intermediate measures, it is possible to calculate $Csec_m$ for the generic secondary measure as follows:

$$Csec_m = Csecmax_m - \left(\sum_{m=1}^{Mpr} Cpr_m * Cint_m * (Csecmax_m - 1) \right)$$

where:

- $Csecmax_m$ is a parameter >1 that can be used to estimate the time to the tunnel closure independently from the measures that are present;
- Mpr is the total number of primary measures;
- Cpr_m is a coefficient associated with each primary measure, variable between 0 and 1. If there is no measure, the value is null. The sum of the values of Cpr_m may reach 1.

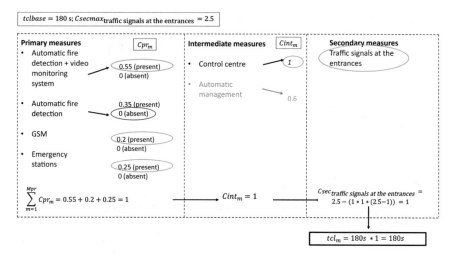

Fig. 4.11 Calculation of tch_m with all the primary measures operational

- $Cint_m$ is a coefficient associated with each intermediate measure variable between 0 and 1.

Some numerical examples that refer to the strategy shown in Fig. 4.8 are illustrated in the following.

Figure 4.11 illustrates a case in which the traffic lights outside the tunnel are activated by three primary measures, among which automatic event identification with cameras, with the control centre as the intermediate measure.

$Csec_m$ is equal to 1: the result is that tcl_m is equal to $tclbase$.

Figure 4.12 shows a case in which the traffic lights outside the tunnel are present but the four primary measures are absent; in addition, the automatic management system, which results as being less performing than the control centre, is also present.

$Csec_m$ is equal to 2.5, and tcl_m would have the same value if the external traffic lights were not present.

An example of a situation in which only some primary measures are present is presented in Fig. 4.13.

The value of $Csec_m$, which is equal to 2.01, is between the best value 1 and the worst value 2.5 ($Csecmax_m$).

4.2 Independent Measures

It must be underlined that the model also includes the existence of some system measures whose operation is independent from that of all the others, such as:

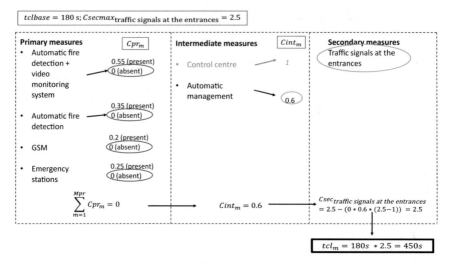

Fig. 4.12 Numerical example for calculating tcl_m with no primary measure present

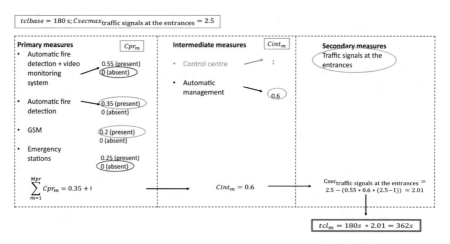

Fig. 4.13 Numerical example for calculating tcl_m with only some measures present

- Road signs;
- Evacuation lighting;
- Emergency walkways.

As the independent measures belong to the secondary class, the coefficient $Csec_m$ can be defined for each of them.

Regarding road signs that influence the user movement speed v_m, for example, if they are present $Csec_{road\ signs} = Csecmax_{road\ signs} (>1)$, if they are not present $Csec_{road\ signs} = 1$. In this manner, multiplying $Csec_m$ by the basic movement speed $vbase$ it is possible to consider the presence or not of the system measure.

Chapter 5
Queue Formation Model

Abstract This chapter describes in detail the vehicle queue formation model for each lane inside the tunnel following an accident.

The model gives the instantaneous length along which the queue of vehicles extends in each *i-th* lane from the accident location to the tunnel entrance.

Estimation of the queue length is needed to determine the number of users potentially involved in the event, thus it is necessary to know the number of vehicles, their position and size, and the respective occupants.

5.1 Queue Formation Speed

The model for the queue formation in each *i-th* lane is based on analytical vehicle flow formulations that relate flow, speed and density, from which the formula for the kinematic speed of queue formation can be determined as follows:

$$u_i = \frac{QB_i - QA_i}{KB_i - KA_i}$$

where:

- u_i represents the queue formation speed in the *i-th* lane;
- QA_i and QB_i represent respectively the flow of vehicles in the interrupted flow (post event) and free flow (before the event) conditions in the *i-th* lane and are expressed as vehicles/h;
- KA_i and KB_i are the vehicle densities in the interrupted flow (post event) and free flow (before the event) conditions in the *i-th* lane and are expressed as number of vehicles/km;

As an example, Fig. 5.1 shows a layout that represents queue progress following a traffic disturbance (event).

© The Author(s) 2019
F. Borghetti et al., *Road Tunnels*, PoliMI SpringerBriefs,
https://doi.org/10.1007/978-3-030-00569-6_5

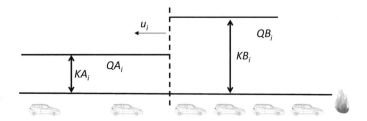

Fig. 5.1 Progress of the queue front after an event in the *i-th* lane

In Fig. 5.1 it can be seen that the incoming vehicles on the left still move in a free flow condition, while the vehicles on the right are stopped in a queue following an accident.

If we consider:

- $QA_i = Q_i$;
- $QB_i = 0$;
- $KB_i = Dqueue_i$;
- $KA_i = \frac{QA_i}{vA_i}$;
- $vA_i = vFL_i$;

we obtain:

$$u_i = \frac{QB_i - QA_i}{KB_i - KA_i} = \frac{-Q_i}{Dqueue_i - \frac{Q_i}{vFL_i}}$$

The difference between the density of the vehicles in the queue $Dqueue_i$ and the density in the free flow condition $\frac{Q_i}{vFL_i}$, is certainly positive because the number of vehicles in a section when stopped in a queue is larger than the number observed if the vehicles had been moving. This implies a negative speed, which means that the queue is forming towards the tunnel entrance (opposite to the traffic direction).

u_i gives an idea of the rapidity of the queue formation. The faster the phenomenon, the longer the queue inside the tunnel and so the greater the number of potentially exposed users.

5.1.1 Time Required for the Queue to Reach the Tunnel Entrance

After having determined the queue formation speed for the *i-th* lane, it is possible to calculate the time needed for the queue to reach the tunnel entrance, therefore saturating the whole section L_p between the point at which the initial event occurred and the entrance, as shown in Fig. 5.2.

L_p is given by the formula:

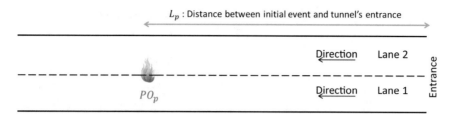

Fig. 5.2 Representation of the parameter L_p as a function of the initial event position PO_p

$$L_p = Ltot * PO_p$$

The analytical formula for calculating the tunnel saturation time is:

$$tsat_{i,p} = \frac{L_p}{u_i}$$

where:

- $tsat_{i,p}$ is the time needed for complete saturation of the i-th lane upstream of the accident;
- L_p identifies the distance between the accident location and the tunnel entrance;
- u_i is the queue formation speed in the i-th lane.

5.1.2   Tunnel Closing Time

The tunnel closing time tcl_m represents the time interval that passes from the occurrence of the initial event until the safety device (e.g. traffic light or VMP) positioned at the tunnel entrance indicates tunnel closing.

The closing time is a parameter of the model that depends on the tunnel system measures MI_m: the absence of one or more measures leads to a higher tcl_m, therefore worsening of the overall situation.

Considering the tunnel closing time tch_m and the queue formation speed u_i, it is possible to determine the length of the queue $Lqcltot_{i,m}$ as follows:

$$Lqcltot_{i,m} = Lqcl1_{i,m} + Lqcl2_{i,m}$$

where:

- $Lqcltot_{i,m}$ is the total length of the queue that can be observed in the i-th lane considering a tunnel closing time of tch_m;
- $Lqcl1_{i,m}$ is the section of queue that has formed inside the tunnel before closure, namely within time tcl_m; the result is that $Lqcl1_{i,m} = u_i * tcl_m$, with u_i representing the queue formation speed in the i-th lane and tcl_m representing the time interval

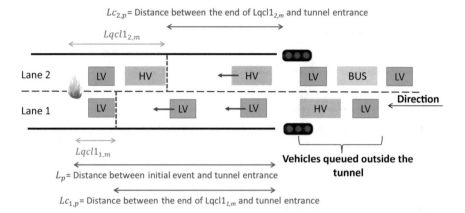

Fig. 5.3 Section of queue $Lqcl1_{i,m}$ and formation of section $Lqcl2_{i,m}$

that passes from the occurrence of the initial event (fire or DG release) until the device (e.g. traffic light or arrow-cross panel) positioned at the tunnel entrance indicates closure to the incoming users;

- $Lqcl2_{i,m}$ is the section of queue formed by the vehicles that entered the tunnel before the closing time tcl_m and which are added to the vehicles that have already stopped. The example given in Fig. 5.3 shows the queue $Lqcl1_{i,m}$ that has formed over the time period tcl_m and the vehicles moving inside the tunnel that will form the section of queue $Lqcl2_{i,m}$. This parameter is especially important for long tunnels.

$Lqcl2_{i,m}$ is obtained from the formula:

$$Lqcl2_{i,m} = \frac{Q_i}{v_i} * \frac{Lc_{i,p}}{Dqueue_i}$$

where:

- Q_i is the vehicle flow in the free flow condition (before the event occurs) in the *i-th* lane and is expressed in vehicles/h;
- v_i is the flow speed of the *i-th* lane given by the average of the speeds of the light and heavy vehicles and buses vLV, vHV, $vBUS$, weighted with the percentages of light and heavy vehicles and buses for the *i-th* lane $\%LV_i$, $\%HV_i$, $\%BUS_i$;
- $Lc_{i,p}$ is the section of tunnel that goes from the end of the queue section $Lqcl1_{i,m}$ to the tunnel entrance, given by $Lc_{i,p} = L_p - Lqcl1_{i,m}$;
- $Dqueue_i$ is the density of stopped vehicles in the *i-th* lane expressed as vehicles/km.

The density of the vehicles that are stopped in the queue can be determined with the formula:

$$Dqueue_i = (1000 + dvehic)/(lLV * \%LV_i + lHV * \%HV_i$$

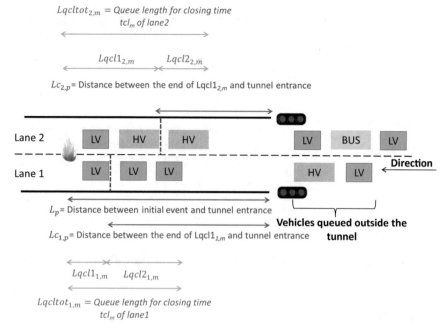

Fig. 5.4   Total length of the queue $Lqcltot_{i,m}$ in the two lanes of the tunnel

$$+ lBUS * \%BUS_i + dvehic)$$

where:

- $dveic$ is the average safety distance between the vehicles stopped in the queue;
- lLV, lHV, $lBUS$ are respectively the average lengths of the light vehicles, heavy vehicles and buses.

Figure 5.4 shows $Lqcltot_{i,m}$ and its two components, $Lqcl1_{i,m}$ and $Lqcl2_{i,m}$

5.1.3   Distance Between the Vehicles Stopped in the Queue

The average distance between the vehicles that are stopped in the queue $dvehic$ varies according to the average daily traffic: if the ADT value is high, the distance between vehicles will be smaller than if the ADT value was low and with a larger distance between vehicles. Consequently, a high ADT leads to a higher number of potentially exposed users in the tunnel, while a low ADT results in a lower number of users stopped in the queue.

A linear dependence between ADT and the average safety distance was defined in the model, as shown in Table 5.1.

Table 5.1 Linear dependence between the ADT and the average safety distance *dvehic*

ADT (vehicles/day)	Average safety distance between the vehicles stopped in a queue *dvehic* (m)
22,000	1
4000	10

Table 5.2 Numeric example of the average safety distance *dvehic* according to the ADT

ADT (vehicles/day)	Average safety distance between the vehicles stopped in a queue *dvehic* (m)
18,818	3
13,421	5
6710	8

As an example, Table 5.2 shows the value of *dvehic* for three different ADT values.

5.1.4 Queue Length

One of the main hypotheses on which the model is based consists of the fact that the vehicles downstream of the event, compared to the running direction, are free to continue towards the tunnel exit, moving quickly to a safe place, while the vehicles upstream of the event, being stopped in a queue, are potentially exposed to the consequences of an accident and are therefore involved in the evacuation process.

In the case of long closing times, the queue can extend beyond the tunnel entrance.

The next step consists of comparing the tunnel closing time tcl_m with the time needed to completely saturate the *i-th* lane $tsat_{i,p}$ defined previously. In this manner, it is possible to know the effective length of the queue in each *i-th* lane.

Two distinct situations can be obtained, described as follows:

- $tsat_{i,p} \leq tcl_m$
- $tsat_{i,p} > tcl_m$.

5.1.5 Case 1: Queue Length Longer Than the Upstream Tunnel Section

In the case where $tsat_{i,p} \leq tcl_m$ the tunnel saturation time is lower than the tunnel closing time, therefore the tunnel fills with vehicles before the external devices inform the incoming users that the tunnel is cloded. The queue certainly occupies the length L_p and above all could extend in the open air for a length equal to $Lqcltot_{i,m} - L_p$, as illustrated in Fig. 5.5.

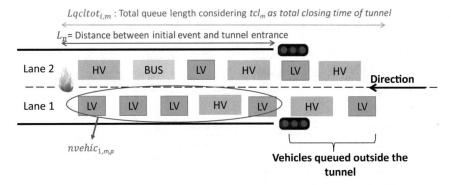

Fig. 5.5 Representation of total vehicle saturation inside the tunnel

As the vehicles in the queue outside the tunnel are not considered for calculation purposes because they are not subject to the potential dangers that exist in a confined environment, the queue length becomes:

$$Lqueue_{i,m,p} = L_p$$

The result is that, in case 1, the number of vehicles involved in the *i-th* lane is determined as follows:

$$nvehic_{i,m,p} = L_p * Dqueue_i$$

where:

- $nvehic_i$ is the number of vehicles involved by an accident in the *i-th* lane;
- $Dqueue_i$ is the density of vehicles stopped in the *i-th* lane;
- L_p is the distance between the accident and the tunnel entrance.

5.1.6 Case 2: Queue Length Shorter Than the Upstream Tunnel Section

If, instead, $tsat_{i,p} > tcl_m$, the saturation time is larger than the tunnel closing time, therefore the devices located outside inform the incoming users that the tunnel is closed, preventing additional vehicles from entering and filling the tunnel.

In this second case, the queue does not fill the whole length between the accident position and the tunnel entrance L_p but becomes $Lqueue_{i,m,p} = Lqcltot_{i,m} = Lqcl1_{i,m} + Lqcl2_{i,m}$, as shown in Fig. 5.6.

The result is that the number of vehicles involved in the *i-th* lane is determined as follows:

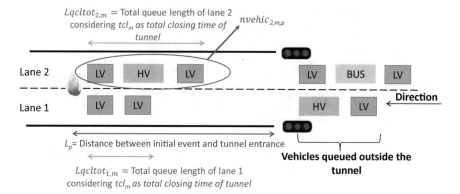

Fig. 5.6 Representation of partial tunnel filling

$$nvehic_{i,m,p} = Lqcltot_{i,m} * Dqueue_i$$

The queue formation model for case 2 identifies the number of vehicles in the queue in the *i-th* lane as:

$$nvehic_{i,m,p} = Lqueue_{i,m,p} * Dqueue_i$$

where:

- $Lqueue_{i,m,p} = \begin{cases} L_p & if\ tsat_{i,p} \leq tcl_m \\ Lqcltot_{i,m} & if\ tsat_{i,p} > tcl_m \end{cases}$
- L_p is the distance between the tunnel entrance and the accidental event position;
- $Lqcltot_{i,m}$ is the extension reached by the queue of vehicles in the *i-th* lane after the closure of the tunnel;
- $Dqueue_i$ is the density of the vehicles stopped in the queue in the *i-th* lane.

5.1.7   Number of Potentially Exposed Tunnel Users

To estimate the number of users who are potentially exposed to the event, the division between the specific categories of vehicle must be known and each one attributed with an average occupation coefficient.

The sum of the number of users calculated for the various lanes gives the total number of the individuals that are present simultaneously inside the tunnel.

The number of users in the *i-th* lane results as being expressed by:

$$N_{i,m,p} = nvehic_{i,m,p} * (\%LV_i * OLV + \%HV_i * OHV + \%BUS_i * OBUS)$$

where:

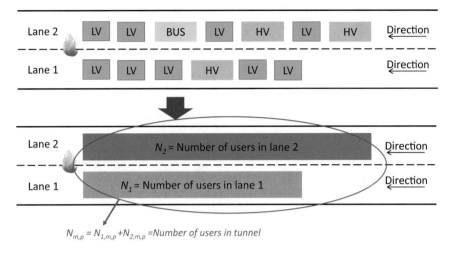

$N_{m,p} = N_{1,m,p} + N_{2,m,p} = Number\ of\ users\ in\ tunnel$

Fig. 5.7 Representation of the users present $N_{m,p}$ in lanes 1 and 2

- $\%LV_i$, $\%HV_i$, $\%BUS_i$ are the percentage of light vehicles, heavy vehicles and buses respectively in the *i-th* lane;
- *OLV*, *OHV*, *OBUS* are the average occupation coefficients respectively of light vehicles, heavy vehicles and buses in the *i-th* lane.

If, on the one hand, it can be confirmed that the vehicle occupation coefficient is a personal characteristic of the vehicle itself and can be considered equal for each lane, the percentage of specific categories of vehicles is different from one lane to the other. As an example, fast lanes tend to have a larger presence of light vehicles than heavy ones.

The total number of users present inside the tunnel during the accidental event can be obtained by summing the users present in each *i-th* lane, as shown in Fig. 5.7.

$$N_{m,p} = \sum_{i=1}^{I} N_{i,m,p} = \sum nvehic_{i,m,p} * (\%LV_i * OLV + \%HV_i * OHV$$
$$+ \%BUS_i * OBUS)$$

where *I* is the total number of lanes in the tunnel.

As an example, Fig. 5.7 shows a representation of the number of users for each lane starting from the number of light vehicles, heavy vehicles and buses.

Chapter 6
Distribution Model of Potentially Exposed Users

Abstract This chapter describes the distribution model of users in vehicles that are stopped in a queue in each lane of the tunnel. The model allows to estimate the number of users present along the queue of each lane according to the traffic and the type of vehicles.

After having estimated the number of users present in the tunnel and potentially exposed to the event, the purpose of the distribution model is to divide the vehicle occupants along the whole extension of the queue for each i-th lane.

The model used is based on a uniform distribution of the users in each i-th lane because the queue length can be different in the individual lanes and an equal distribution of vehicles in all the lanes would not guarantee a correct representation of the true position of the vehicles and their occupants.

The users are positioned in a uniform and homogeneous manner along the queue. The density of users is equal to:

$$dul_{i,m,p} = \frac{N_{i,m,p}}{Lqueue_{i,m,p}}$$

where:

- $N_{i,m,p}$ is the number of potentially exposed users in the i-th lane;
- $Lqueue_{i,m,p}$ is the extension in metres that the queue reaches upstream of the event in the i-th lane and which depends on the queue formation model.

This uniform distribution method requires the introduction of lane discretization using cells.

The cell, therefore, represents the minimum element into which the queue length is divided for the i-th lane. The users belonging to the cell are concentrated in its geometrical centre: in this manner a discrete point analysis, not a point-by-point one, is carried out in order to find a good compromise between the computational burden and the accuracy of the result. It is evident that both the computational burden and calculation precision strongly depend on the length lc of the cell.

The number of users in the j-th cell located in the i-th lane is determined by:

© The Author(s) 2019
F. Borghetti et al., *Road Tunnels*, PoliMI SpringerBriefs,
https://doi.org/10.1007/978-3-030-00569-6_6

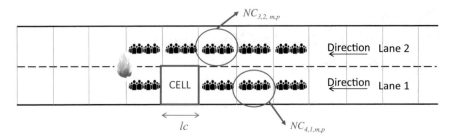

Fig. 6.1 Representation of discretization in cells and the respective associated users $NC_{i,j,m,p}$

$$NC_{i,j,m,p} = dul_{i,m,p} * lc$$

where:

- $dul_{i,m,p}$ is the number of users per unit of length distributed evenly along the queue in the i-th lane;
- lc is the cell length.

Figure 6.1 shows the grid obtained using cells whose width is equal to the dimension of the lane and which have a variable length equal to lc.

The length of the cell, lc, is individual for all the tunnel lanes, no matter what the distance from the accident is, while the number of users associated with each cell, $NC_{i,j,m,p}$, varies from lane to lane because the uniform density, $dul_{i,m,p}$, can be different.

Each lane is therefore characterised by a number of cells, $NCT_{i,m,p}$, all of the same length, lc, which is such as to reach the length along which the queue extends. The first cell is placed in the tunnel position where the event occurred, as shown in Fig. 6.2.

6.1 Evaluation of the Evacuation Distance

After having illustrated the method through which it is possible to distribute the users present in the tunnel following an accident, the next operation consists in simulating the evacuation process of those individuals who, starting from their initial position, choose the route to follow in order to move towards a place that is considered to be safe.

This phase is delicate because of the unpredictability associated with human behaviour. The elements to be considered in this case involve, for example, the propensity to exit the vehicle, familiarity with the tunnel, the preparedness, in conditions of emergency and extreme danger, physical ability, interaction with other individuals, etc.

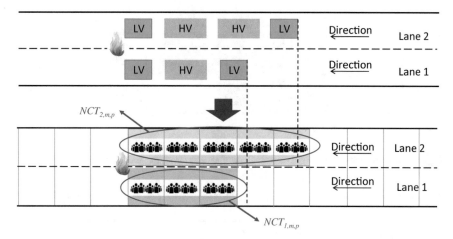

Fig. 6.2 Identification of the cells through which the queue in each lane is discretized

In the proposed model, it was chosen to simulate individual behaviour regarding the choice of the shortest route: it is presumed that the user behaves in compliance with the logic of proximity, opting to reach the closest available exit while moving away from the accidental event (upstream), with a possible exception described in detail in the next paragraphs.

This technique is the simplest and is of general validity, but it presents some gaps for accidental scenarios where the tunnel environmental conditions make some emergency exits unusable: just think, for example, about a high concentration of smoke that could make the emergency exits difficult to see during evacuation, or ineffective emergency signs, or cases in which particular behaviour has to be simulated because the user is not familiar with the confined environment (this can lead a user to escape towards the tunnel entrance without considering that there are closer emergency exits along the path).

In these cases, the choice process of the user can be complex when compared with proximity logic because other variables could influence the decision-making process.

The escape route consists of the distance that each group of users associated with the same discretization cell has to cover in order to move to a safe place, by reaching the tunnel entrance or by using the available emergency exits and/or by passes.

The method analyses the position of the barycentre $G_{i,j}$ of each j-th cell situated in the i-th lane, introducing a distance $xG_{i,j}$ between the barycentre of the cell and the section involved in the accidental event. This distance is determined on the x axis, which originates from the event and is parallel to the longitudinal axes of the tunnel, as shown in Fig. 6.3.

If we consider the i-th lane, the distance between the accidental event and the barycentre of the first cell is:

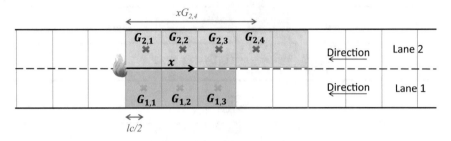

Fig. 6.3 Representation of the barycentre $G_{i,j}$ of the cells along the x axis

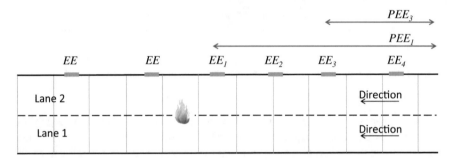

Fig. 6.4 Representation of the emergency exits and the respective distances to the tunnel entrance

$$xG_{i,1} = \frac{lc}{2}$$

The distance between the accidental event and the barycentre of the second cell is:

$$xG_{i,2} = xG_{i,1} + lc$$

and so on progressively

$$xG_{i,NCT_{i,m,p}} = xG_{i,\left(NCT_{i,m,p}-1\right)} + lc$$

where $NCT_{i,m,p}$ is the total number of cells relative to the n-th lane into which the queue is divided.

The model considers the presence and position of each emergency exit/bypass, EE_k, and the position of the tunnel entrance as shown in Fig. 6.4. The progressive PEE_k with respect to the tunnel entrance is therefore defined for each k-th emergency exit.

A fundamental hypothesis of the implemented model is that a user, while choosing the escape route, considers the first available upstream exit or the tunnel entrance. This hypothesis is justified by the fact that in emergency conditions and with panic,

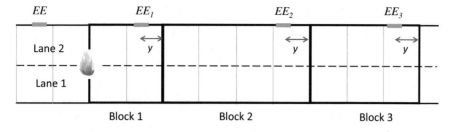

Fig. 6.5 Representation of the threshold y that defines each cell block

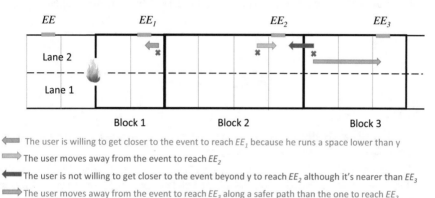

Fig. 6.6 Possible escape routes that can be used by users as a function of the initial position (cell)

users generally tend to step away from the accidental scenario where the highest concentration of smoke and toxic gas is, and the presence of flames can create a dangerous situation.

An exception to this general criterion is the case in which the closest emergency exit is downstream (in the running direction) and not upstream within a threshold y, measured from the cell barycentre to the emergency exit.

This exception can be confirmed by the fact that users are willing to move towards the accident position if the distance covered, y, is short; in this way he reduces the distance needed to reach the closest upstream exit.

The parameter y therefore identifies blocks of cells to which a single emergency exit EE_k can be associated, as shown in Fig. 6.5. It is evident that users associated with the cells located beyond the last emergency exit upstream will use the tunnel entrance to escape.

Figure 6.6 shows some representative cases that present the position of the cell barycentre $G_{i,j}$ and the choice of emergency exit EE_k as a function of the blocks and the y parameter.

The next step with the model is calculating the distance covered between the barycentre of each cell $G_{i,j}$ and the emergency exit EE_k that is available (belonging to the same block). This distance is determined by evaluating the longitudinal and transversal contributions of the path:

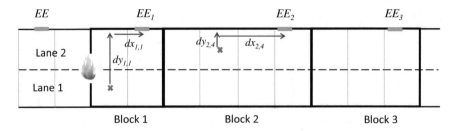

Fig. 6.7 Escape route of users according to the initial position (cell barycentre). The figure also shows the distance to be covered, given by the longitudinal and the transversal contributions

$$dcell_{i,j} = dx_{i,j} + dy_{i,j}$$

where:

- $dx_{i,j}$ is the contribution of the longitudinal distance associated with the barycentre of the *j-th* cell belonging to the *i-th* lane;
- $dy_{i,j}$ represents the contribution of the transversal distance associated with the barycentre of the *j-th* cell belonging to the *i-th* lane and, as a general rule, that can be assumed as being equal to half the lane width.

Figure 6.7 illustrates an example of the distance $dcell_{i,j}$ between the generic cell and the first emergency exit available, with particular reference to the longitudinal $dx_{i,j}$ and transversal $dy_{i,j}$ contributions.

The choice to algebraically sum the two contributions, longitudinal and transversal to the tunnel axis, and not calculate the diagonal route that connects the barycentre of the cell with the emergency exit can be justified as follows:

- the presence of the stopped vehicles in the queue could in some cases make it physically impossible to cover the diagonal route;
- it was considered as reasonable to represent human behaviour as tending to immediately move towards the side where emergency exits and other devices (optical and/or physical) are present and can support the egress of the user.

The final result of the uniform distribution method consists of the overall distance that each individual belonging to a cell has to cover in order to reach the first available emergency exit. The shorter the cell length, the more the model tends to represent the effective distance covered by each user present into the tunnel, considering the real position.

It is evident that using cells characterised by a very limited size, in addition to increasing the calculation times, will result in a simulation with an excessive discretization that possibly does not mirror reality.

Chapter 7
Consequence Analysis of the Accidental Scenarios

Abstract This chapter describes the approach used to determine the evolution of the consequences for each of the accidental scenarios considered in the risk analysis procedure. The consequence analysis allows to estimate which are the negative effects of the accidents that can affect both the egress and tenability of tunnel users.

After having defined which are the reference accidental scenarios to be considered in the risk analysis procedure, as mentioned in the previous chapters, the possible negative effects and their evolution in time and space within the tunnel due to these accidents have to be estimated in order to determine which accidental scenarios are able to limit the egress and tenability of the tunnel users.

The quantitative consequence analysis of the accidental scenarios allows to determine several analytical correlations that have to be inserted in the tunnel risk analysis model for the estimation of *Tfs*, which is the progress time of the effects of the *s-th* accidental scenario; in other words, this analytical process is used to evaluate the ASET (Available Safe Egress Time), time to untenable conditions. Safety is defined by an assessment based on a set of acceptance criteria which are basically expressed in terms of tenability limits to occupants who may be potentially exposed to the effects of one or more accidents [1, 2]. According to the considered accidental scenario, a set of acceptance criteria for tenability limits can be considered; in particular, for fire events and dangerous goods releases, tenability thresholds can be defined for the following common parameters:

- thermal radiation;
- smoke temperature;
- toxicity of gas and/or vapors;
- overpressure;
- visibility in smoke.

The above define the safety requirements needed to estimate the ASET; among them, visibility affects also the evaluation of the RSET (Required Safe Egress Time), as will be described in the following chapter.

Apart from the tenability criteria, the ASET determination rely upon a number of features of the *s-th* accidental scenario, such as the source term, the tunnel geometry and emergency equipment performances. As a consequence, the effects of every

© The Author(s) 2019   55
F. Borghetti et al., *Road Tunnels*, PoliMI SpringerBriefs,
https://doi.org/10.1007/978-3-030-00569-6_7

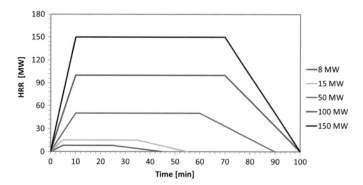

Fig. 7.1 Example of fire curves that can be used in road tunnel risk analysis

accidental scenario have a slow or fast propagation inside the tunnel as a function of the fire intensity or the severity of the DG release considered. As an example, a fire curve, together with combustion products characteristic (at least, in terms of smoke yield), must be defined to quantitatively analyze the fire consequences; moreover the consequence analysis model have to consider the interaction between fire and the tunnel geometry, as well as the role of smoke control systems or fire suppression systems. Considering typical HRR values that can be associated with fires of different types of vehicles (Table 3.1 and Fig. 3.7), it is possible to design fire curves has those reported in Fig. 7.1.

Several theoretical or modelling approaches can be used to quantitatively evaluate and describe the relevant negative effects of a possible accidental scenario and how they propagate inside the road tunnel. As an example, a preliminary and rough estimation of the consequences of fires or DG releases can be obtained by means of theoretical or empirical correlations; on the other hand, in order to have a better representation of some complex accidents, a more detailed analysis can be performed by using integral models or CFD codes [3–11].

7.1 Quantification of the Consequences of the Accidents

The field of computational modeling has had important advances since the 1980s, due to progress in understanding the physics of phenomena such as fire, the availability of complete and accurate experimental data, the formulation of more detailed models and the access to powerful computational resources with increased process capacity [10]. ISO/TR 13387 distinguishes numerical codes used in fire modeling in two categories of different complexity: zone models and field models [12].

7.1.1 Computational Fluid Dynamics for Fire Safety and Accident Analysis

The fundamental role of numerical modeling in Quantitative Risk Analysis (QRA) and Fire Safety Engineering (FSE) has been broadly accepted in the scientific community [13]. As previously mentioned, for a realistic numerical prediction of fire, useful for consequences quantification and risk assessment, the complex interaction between turbulent flow, buoyancy, convection, air entrainment, non-premixed combustion, soot formation, thermal radiation, fluid-structure interaction, dispersion of smoke and toxic combustion products and the effects of possible mitigation barriers or emergency measures need to be modeled. In order to achieve so, a reliable software tool must take into account the fundamental physics and complex chemistry behind an uncontrolled combustion phenomenon and provide results in a reasonable amount of time.

Computational fluid dynamics (CFD) codes solve the conservation equations describing the transport of mass, momentum, energy and species to analyze the evolution of fluid flows. As an example, the equations are treated so that they can be efficiently solved for the fire scenario of interest; the physical space, in which the flow field needs to be solved, is typically divided in small control volumes (mesh) and the governing equations are integrated over each volume. By numerically solving these discretized conservation equations and adopting several sub-models, CFD codes can have a very high degree of accuracy, so that amongst their advantages there is the possibility to study complex systems or hazardous scenarios. However, a complete CFD model of the dynamics of fires must account for the mutual interactions of fire physical and chemical effects, namely, fluid mechanics, turbulence, heat transfer, combustion chemistry, materials and geometry. For this reason, the codes must be carefully tested for different classes of scenarios in order to ensure that appropriate results are predicted and in order to identify potential limitations. Testing the model under known conditions and comparing the simulated results to data from a particular reference test is important before using the model to predict the results of a similar, never tested scenario [14]. It is important to highlight that in a real tunnel fire, the heat release rate varies with time. At the same time, the ventilation velocity across the fire site also varies with time due to the intervention of the emergency ventilation system just after the fire is detected. Therefore, the fire behavior is highly dynamic and some characteristic parameters such as the flame angle, back-layering length and maximum smoke temperature beneath the ceiling become transient values. Despite the fact that the behavior of a real tunnel fire is highly dynamic, CFD codes, such as FDS, are able to capture with satisfactory agreement the complex behavior of transient full scale tunnel fire tests [4].

About fire scenarios, the estimation of their effects can be been performed with a computational fluid dynamics code, such as the Fire Dynamics Simulator (FDS) from the American National Institute of Standards and Technologies (NIST), a code used for low-speed, thermally driven flows with an emphasis on heat transport and smoke movement. FDS [15] solves a form of the Navier-Stokes equations for weakly

Fig. 7.2 Dynamics of the smoke spread inside a road tunnel (length of about 2000 m, cross-section of 55 m^2) predicted by FDS for a fire with a maximum HRR of 50 MW: **a** no mechanical ventilation; **b** with emergency ventilation; the white arrows represent the direction of the natural airflow that affects the fire and smoke evolution

compressible reacting flows travelling at relatively low speed compared to the speed of sound. The Rehm and Baum simplified equations, or "low Mach number combustion equations", describe the low speed motion of a gas driven by chemical heat release and buoyancy forces [16], for which is possible to neglect the contribution of acoustic pressure waves in the balance. The partial derivatives of the conservation equations of mass, momentum and energy are approximated as finite differences, and the solution is updated in time on a three-dimensional, rectilinear grid. In fires, soot is usually the most important source and absorber of radiation; therefore the modeling of soot formation and oxidation processes is therefore important for the accurate prediction of radiation emissions and visibility reduction; the version used relied upon a user-specified soot yield for the combustion modeling, while the radiation scattering from gaseous species and soot is not included in the model [17].

As reported in the example of Fig. 7.2, FDS can easily reproduce a transient fire scenario, considering the influence of both natural and mechanical ventilation on the smoke spread and visibility reduction inside a two-lane road tunnel. This quantitative and detailed modeling approach allows also to evaluate the possible influence of the location of the accidental event on the dynamics of fire effects; without specific information, it is usually assumed that the source of the accidental event can be located in a generic position of the tunnel, defined by a percentage with respect to the total length of the tunnel; for this reason, ad already mentioned, the proposed

Fig. 7.3 Effect of the fire source location on the thermal effects propagation inside the road tunnel for two fire scenarios of Fig. 7.1: **a** HRR = 8 MW; **b** HRR = 150 MW. A temperature threshold of 90 °C was used as tenability criterion to obtain the curves for the ASET calculation; the direct comparison among these curves is highlighted in (**c**) and (**d**) for the considered scenarios

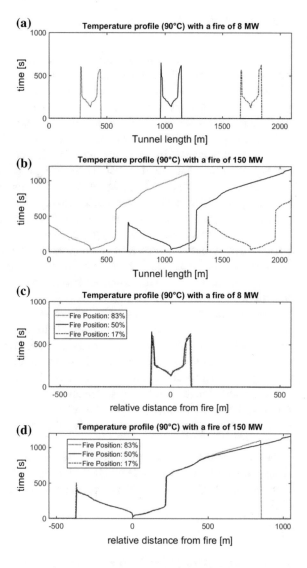

risk analysis model considers for each traffic direction at least 6 different positions in which the accidental scenarios may occur. CFD simulation of some reference fire scenarios allowed to demonstrate that the propagation of the fire consequences, for tunnels with homogeneous characteristic, is basically independent of the fire location within the tunnel, thus strongly reducing the computational costs associated to the detailed modeling of all the considered scenarios (Fig. 7.3).

7.2   Zone Models

Under certain conditions, fire effluents in a compartment tend to stratify in a hot smoke layer in proximity of the ceiling, leaving a cold layer of fresh air below.

Zone models are one-dimensional models that rely on the assumption that a compartment volume can be vertically subdivided into two perfectly mixed zones with homogeneous properties in terms of temperature and composition: a hot layer with combustion products located near the ceiling, and a cold layer with fresh clean air at the bottom, separated by a moving interface. The properties (and the layer height) can vary over time and are identified by solving global conservation equations.

Each zone (cold and hot layer, fire plume and compartment boundary) is modeled separately and then linked to the others through fluid dynamic and heat transfer equations. The fire plume acts as a pump that transfers mass and heat between zones. This approach simplifies the computations, as each zone is treated separately. Experimentally-based expressions can be used to describe heat and mass transport processes among the zones. The output from one zone is used as the input to another, and sub-models (such as the ceiling-jet correlations or plume correlations) can be incorporated in the model.

For this reason, zone models rely on very strong simplifications as well as on empirical correlations, and their application is limited to the geometry characteristics over which each model was tested and validated. They can be applied with a good degree of confidence to relatively simple enclosure fires, as long as detailed spatial distribution of physical properties is not required and the two-layer description reasonably approximates reality. These models are commonly used in risk analysis of rooms and simple buildings [18], and given the little computational time required for each run—usually in the order of seconds—they are considered useful for iterative design [19].

When dealing with a complex three-dimensional flow of gases and hot products rising from a fire located inside a tunnel, this cannot be reproduced directly with a two-zone approach using a single compartment to represent the whole tunnel volume [19].

7.2.1   Zone Model Software for Tunnel Fires

A variety of software packages for ventilation simulation of tunnels have been applied to fire safety problems [20]. These are generally based on a one-dimensional view of the fluid flow (i.e. a single value of pressure, temperature and velocity represents the entire flow at each cross-section) and as such cannot represent stratification, backflow or near field.

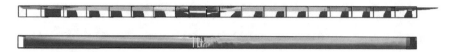

Fig. 7.4 Temperature distribution inside the tunnel predicted by CFAST (top), with a 19 zone configuration, and FDS (bottom) for the test case Apte_2 with u = 0.85 m/s

CFAST is a two-zone fire model proposed by NIST used to calculate the smoke dispersion, the fire gases dynamics and the temperature throughout compartments of a constructed facility over time. The fundamental equations (conservation of mass and energy over the layers, ideal gas law and relations for density and internal energy) are implemented as system of ordinary differential equations (ODEs), which are solved to give the values of pressure, layer heights and temperatures over time. A series of algorithms allow to compute the mass and enthalpy source terms required by the system [18].

CFAST software has an interface which supports the user in the definition of the scenario, allowing to describe step by step the geometry of the system, the openings of the compartment to the outside and between different compartments, the mechanical ventilation flows, as well as the fire source terms. The code also contains a database of typical fire sources burning rate curves (including emissions) and building materials thermal properties, which can be modified by the user.

A first validation of the CFAST code for tunnel fire analysis is based on the tests performed by Apte et al. [21] inside a mining tunnel with a length of 90 m for different ventilation conditions, with fires characterized by a HRR in the range 2–2.4 MW; the simulation domain in CFAST was realized by dividing the domain into a number of compartments along the tunnel axis, in order to overcome the limitations of the single-compartment approach, which does not allow to describe the smoke destratification along the tunnel length. As an example, the hot gas layer position as well as the temperature distribution predicted by CFAST and FDS are reported in Fig. 7.4 for one of the investigated cases. The comparison shows the capability of CFAST to capture the overall fire behavior but also the impact of the simplifications, which result in differences in the temperature especially in the lower zones.

A comparison among different test cases simulated with the CFAST and FDS is reproduced in Fig. 7.5, where it can be noticed that CFAST achieves reasonable agreement with experimental data especially at low ventilation velocities, when the smoke layer extends over the entire tunnel length. The zone model cannot fully represent the effect of the increase in ventilation velocity, when the smoke backlayering length extends over a short length of tunnel (test Apte_3 with u = 2 m/s). On the other hand, CFAST is able to describe the overall dynamics of the phenomenon and the temperature trend along the tunnel. It does not correctly reproduce the effect of ventilation, as FDS does, but it makes a slightly conservative overestimation of the experimental results.

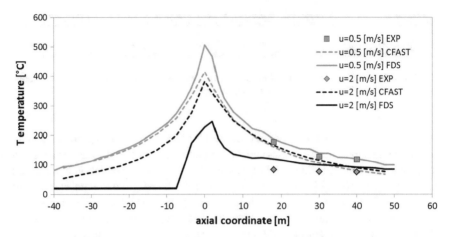

Fig. 7.5 Effect of the variation of the longitudinal ventilation on the predicted temperature profiles in FDS (continuous lines) and CFAST (dashed lines). The experimental data are also included

Given the uncertainty of a typical accidental scenario in tunnel risk analysis, regarding the position, fire curve and maximum HRR, materials, and the variability of traffic conditions (congestion), CFAST can be considered as a reasonable tool for risk analysis; in addition, considering the differences in terms of computational requirements, simulation times and expertise required to perform a fire safety analysis with simulation tools of different levels of approximation, a series of scenarios, which develops commonly over a time scale of 0.5–1 h, can be conveniently simulated and analyzed with CFAST (or other zone models) to evaluate the influence on the smoke and fire dynamics of several parameters such as the fire position, the fire heat release rate, and the ventilation characteristics. The high number of scenarios and tunnels to be considered in a risk analysis often does not allow the use of CFD codes due to the prohibitive calculation burden both in terms of time and performance of the calculators.

As an example, in the following figures some results obtained by means of CFAST are reported for fire scenarios characterized by a maximum HRR of 8 and 50 MW (Figs. 7.6 and 7.8), in which the intervention of the emergency ventilation system is also considered (Figs. 7.7 and 7.9). In both cases there is an initial natural air flow with an air speed of −2.5 m/s, which represents an unfavorable condition for unidirectional tunnels. These maps refer to a section of the tunnel close to the position of the fire, representing the evolution in time and space of some of the parameters of interest in the evaluation of the negative effects due to a fire: smoke temperature in the upper area and lower tunnel, carbon monoxide concentration (CO) and optical density of the smoke (visibility-related parameter) in the lower part of the tunnel. The results clearly underline the progressive development of the fire and of the effects due

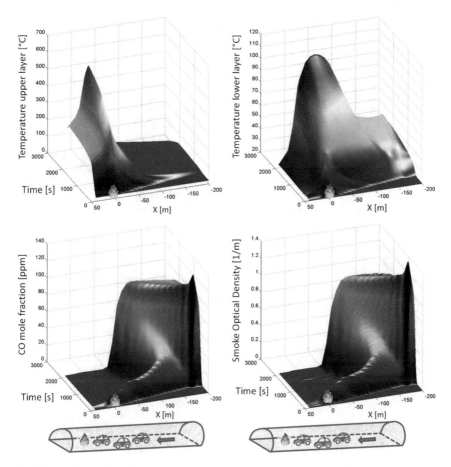

Fig. 7.6 Evolution of smoke temperature in the upper (left) and lower (right) part of the tunnel, carbon monoxide concentration (CO) and optical density of smoke in the lower tunnel area obtained from a CFAST simulation for a Fire with a maximum HRR of 8 MW, without emergency ventilation

to the intervention of the mechanical emergency ventilation system; the fire reaches its maximum power after 600 s from the beginning of the event. Moreover, after about 500 s the piston effect, initially present due to the movement of vehicles, lasts as a result of the formation of the queue of vehicles upstream of the fire and after about 750 s the emergency ventilation begins to counteract the back-layering propagation of the smoke; after about 1200 s the ventilation reaches steady operating conditions and limits the smoke and combustion products presence in a section of the tunnel of a limited length.

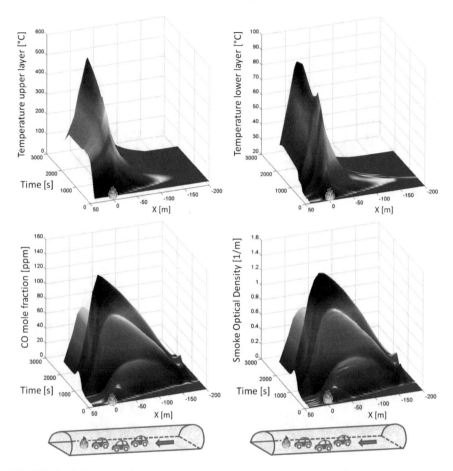

Fig. 7.7 Evolution of smoke temperature in the upper (left) and lower (right) part of the tunnel, carbon monoxide concentration (CO) and optical density of smoke in the lower tunnel area obtained from a CFAST simulation for a Fire with a maximum HRR of 8 MW, with emergency ventilation (activation after 200 s from the beginning of the fire event)

As for CFD results, CFAST simulations of different fire scenarios allows to define typical ASET curves to be inserted into the tunnel risk analysis model, as summarized in Fig. 7.10. Dashed lines represent the dynamics of thermal effects without emergency ventilation, while continuous lines refer to thermal effects evolution when the emergency ventilation system is activated. In particular, this plot clearly highlights the effect of the emergency ventilation activation on the hot smoke propagation upwind from the fire position, which is able to progressively reduce the smoke back-layering and the possible negative thermal effects for the tunnel occupants. The interaction between emergency ventilation and fire smoke is almost negligible for fire scenarios

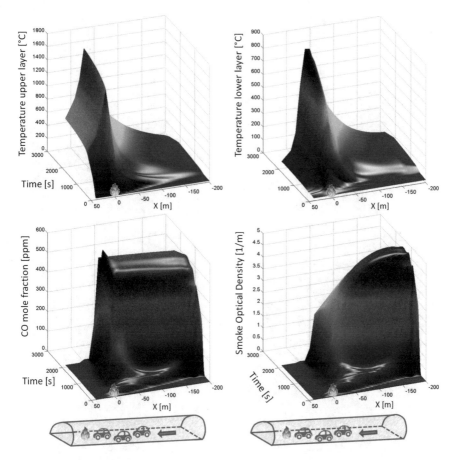

Fig. 7.8 Evolution of smoke temperature in the upper (left) and lower (right) part of the tunnel, carbon monoxide concentration (CO) and optical density of smoke in the lower tunnel area obtained from a CFAST simulation for a Fire with a maximum HRR of 50 MW, without emergency ventilation

characterized by a fast dynamics and a limited HRR, such as the case of a 8 MW fire, because the time required to activate and to stabilize the ventilation is of the same order of magnitude of the time required to the fire to reach steady burning conditions. On the other hand, a much stronger interaction between smoke propagation and mechanical ventilation can be observed for fire with a slow growth rate and a relevant HRR, as the 50 MW fire scenario. These results affect both the ASET and the RSET estimation, as will be discussed in details in the following chapters.

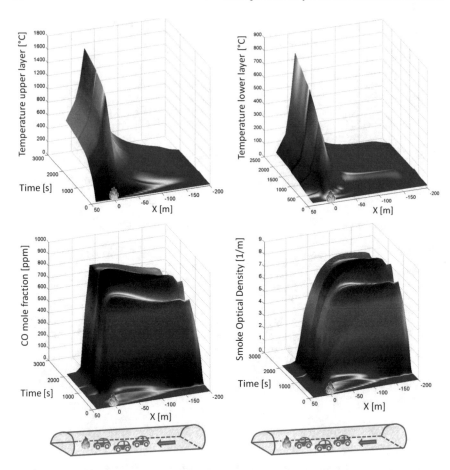

Fig. 7.9 Evolution of smoke temperature in the upper (left) and lower (right) part of the tunnel, carbon monoxide concentration (CO) and optical density of smoke in the lower tunnel area obtained from a CFAST simulation for a Fire with a maximum HRR of 50 MW, with emergency ventilation (activation after 200 s from the beginning of the fire event)

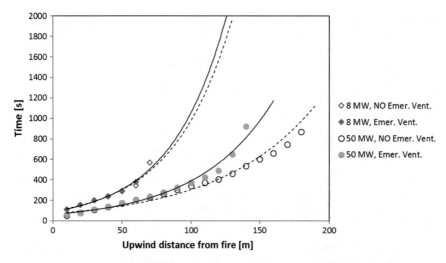

Fig. 7.10 Comparison among different curves obtained with CFAST to describe the maximum upwind extent of thermal effects within the tunnel for the fire scenarios from Figs. 7.6, 7.7, 7.8 and 7.9; a temperature threshold of 90 °C was used as tenability criterion to obtain the curves for the ASET calculation

References

1. Poon SL (2014) A dynamic approach to ASET/RSET assessment in performance based design. Procedia Eng 71:173–181
2. ISO 13571:2012 (2012) Life-threatening components of fire—guidelines for the estimation of time to compromised tenability in fires. International Standardization Organization, Switzerland
3. Favrin S, Busini V, Rota R, Derudi M (2018) Practical les modelling of jet fires: issues and challenges. Chem Eng Trans 67:259–264. https://doi.org/10.3303/CET1867044
4. Tavelli S, Derudi M, Cuoci A, Frassoldati A (2013) Numerical analysis of pool fire consequences in confined environments. Chem Eng Trans 31:127–132. https://doi.org/10.3303/CET1331022
5. Tavelli S, Rota R, Derudi M (2014) A critical comparison between CFD and zone models for the consequence analysis of fires in congested environments. Chem Eng Trans 36:247–252. https://doi.org/10.3303/CET1436042
6. Borghetti F, Derudi M, Gandini P, Frassoldati A, Tavelli S (2017) The test results. In: Tunnel fire testing and modeling: the Morgex north tunnel experiment. Springer International Publishing, pp 39–64. https://doi.org/10.1007/978-3-319-49517-0_4
7. Borghetti F, Derudi M, Gandini P, Frassoldati A, Tavelli S (2017) Evaluation of the consequences on the users safety. In: Tunnel fire testing and modeling: the Morgex north tunnel experiment. Springer International Publishing, pp 65–75. https://doi.org/10.1007/978-3-319-49517-0_5
8. Derudi M, Bovolenta D, Busini V, Rota R (2014) Heavy gas dispersion in presence of large obstacles: Selection of modelling tools. Ind Eng Chem Res 53:9303–9310. https://doi.org/10.1021/ie4034895
9. Di Sabatino S, Buccolieri R, Pulvirenti B, Britter R (2007) Simulations of pollutant dispersion within idealised urban-type geometries with CFD and integral models. Atmos Environ 41:8316–8329

10. Pontiggia M, Derudi M, Alba M, Scaioni M, Rota R (2010) Hazardous gas releases in urban areas: assessment of consequences through CFD modelling. J Hazard Mater 176:589–596. https://doi.org/10.1016/j.jhazmat.2009.11.070
11. Yeoh GH, Yuen KK (2009) Computational fluid dynamics in fire engineering: theory, modelling and practice. Butterworth-Heinemann
12. ISO/TR 13387-3:1999 (1999) Fire safety engineering—Part 3: assessment and verification of mathematical fire models. International Standardization Organization, Switzerland
13. Emmons HW (1985) The needed fire science. In: Proceedings of the first international symposium on fire safety science
14. Kim E, Woycheese JP, Dembsey NA (2008) Fire dynamics simulator (version 4.0) simulation for tunnel fire scenarios with forced, transient, longitudinal ventilation flows. Fire Technol 44(2):137–166
15. McGrattan K, Hostikka S, Floyd J, Baum H, Rehm R, Mell W, McDermott R (2014) Fire dynamics simulator (version 6.1). Technical reference guide—mathematical model. NIST Special Publication 1018-15, US Government Printing Office, Washington, USA
16. Rehm RG, Baum HR (1978) The equations of motion for thermally driven, buoyant flows. J Res NBS 83:297–308
17. Hostikka S (2008) Development of fire simulation models for radiative heat transfer and probabilistic risk assessment. Ph.D. thesis, VTT Publications, Espoo, Finland
18. Floyd J (2002) Comparison of CFAST and FDS for fire simulation with the HDR T51 and T52 Tests. US Department of Commerce, Technology Administration, National Institute of Standards and Technology
19. Meo MG (2009) Modeling of enclosure fires. Ph.D. thesis, University of Salerno, Italy. ISBN 88-7897-032-8
20. Modic J (2003) Fire simulation in road tunnels. Tunn Undergr Space Technol 18(5):525–530
21. Apte VB, Green AR, Kent JH (1991) Pool fire plume flow in a large-scale wind tunnel. In: Fire safety science proceeding of the 3rd international symposium, pp 425–434

Chapter 8
Egress Model of Tunnel Users

Abstract This chapter describes the egress model of the users who are present in the tunnel. Starting from the user position, the available emergency exits and the dynamics of the accidental scenario, it is possible to verify the required safe egress time of the exposed users with the tenability thresholds.

The number of potential users involved in an accidental scenario derives from the queue formation, the distribution of users in the tunnel and the distance to cover in order to reach the closest emergency exit, as illustrated previously.

8.1 Tunnel Zone Model

To estimate the number of potential fatalities, the egress model firstly requires the tunnel to be divided into two zones along the longitudinal axis:

- $z1_s$: is the zone where fatalities occur very quickly (instantaneous) and varies according to the *s-th* accidental scenario;
- $z2_s$: represents the remaining part of the tunnel where the effects of the *s-th* scenario can affect the users. In this zone, the egress process and therefore the users' tenability has to be verified.

Figure 8.1 shows an example of the two zones: the users in the cells located in $z1_s$ are considered as dead because in this zone the acute effects associated with the *s-th* accidental scenario occur very quickly.

8.1.1 Analysis of Zone 1—Z1$_S$

The analysis of $z1_s$ consists of verifying if the barycentre of the *j-th* cell, $G_{i,j}$, is inside this zone: in this case the users inside the cell are considered to be dead.

© The Author(s) 2019
F. Borghetti et al., *Road Tunnels*, PoliMI SpringerBriefs,
https://doi.org/10.1007/978-3-030-00569-6_8

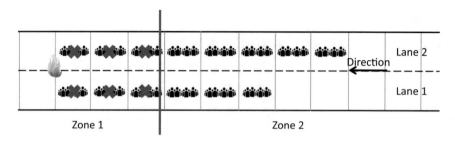

Fig. 8.1 Tunnel zones $z1_s$ and $z2_s$. The users in zone 1 are considered as having died instantaneously

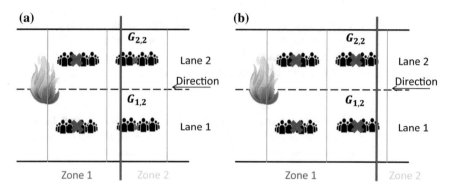

Fig. 8.2 Position of the barycentre of the cells compared to the boundary between $z1_s$ and $z2_s$. In case (**a**) on the left, the barycentre of cells $G_{1,2}$ and $G_{2,2}$ belong to zone 2, while in case (**b**) on the right, the barycentre of cells $G_{1,2}$ and $G_{2,2}$ belong to zone 1

Figure 8.2 shows two examples that demonstrate how in case (a) the barycentre of the cells $G_{1,2}$ and $G_{2,2}$ belong to $z2_s$, while in case (b) they belong to $z1_s$; in this second case the users inside the cells are considered as having died instantaneously.

Continuing, the total number of fatalities in $z1_s$ associated with the s-th scenario can be determined as follows:

$$Ntotfatz1_{m,p,s} = \sum_{i=1}^{I} \sum_{j=1}^{Jz1_i} NC_{i,j,m,p}$$

where:

- I is the total number of vehicle lanes in the tunnel;
- $Jz1_i$ is the number of cells in $z1_s$ for the single i-th lane;
- $Ntotfatz1_{m,p,s}$ is the total number of fatalities in $z1_s$, given by the sum of the users present in the cells belonging to this zone for the s-th scenario;
- $NC_{i,j,m,p}$ is the number of users present in the j-th cell located in the i-th lane.

8.1.2   Analysis of Zone 2—$Z2_S$

When analysing $z2_s$, the model considers the time starting after the occurrence of the accidental event; after this, the egress time of the users in the *j-th* cell of each lane $Tevac_{i,j,m,p}$ should be compared with the time related to the propagation of the effects of each accidental scenario Tf_s.

8.2   Egress Time

The evacuation time of the *j-th* cell is given by the following contributions:

$$Tevac_{i,j,m,p} = Tpr_{i,j,m,p} + Tmov_{i,j,m}$$

where:

- $Tpr_{i,j,m}$ is the pre-movement time, namely the time that passes before the user begins moving towards the closest emergency exit, in other words the evacuation;
- $Tmov_{i,j,m}$ is the movement time, in other words the time required to complete the evacuation.

8.2.1   Pre-movement Time

More in detail, the pre-movement time $Tpr_{i,j,m,p}$ is composed of:

$$Tpr_{i,j,m,p} = Tfc_{i,j,m,p} + Trec_m + Tr_m + Tuvehic_i$$

where:

- $Tfc_{i,j,m}$ is the queue formation time, namely the time the users take to stop in the queue: it increases linearly as the distance from the accident increases, depends on the queue formation speed in each lane and is influenced by the tunnel infrastructure, equipment and management measures;
- $Trec_m$ is the initial event recognition time, meaning the time that users need to recognise the danger situation, and it is influenced by the tunnel infrastructure, equipment and management measures;
- Tr_m is the response time, being the time that passes from the recognition of the accidental event and the beginning of the evacuation from the vehicle. It is influenced by the tunnel infrastructure, equipment and management measures;
- $Tuvehic_i$ is the time taken by the users to exit their vehicles and it is given by the weighted average of the time required to exit a light vehicle, heavy vehicle and bus as regards the composition of the traffic in the *i-th* lane.

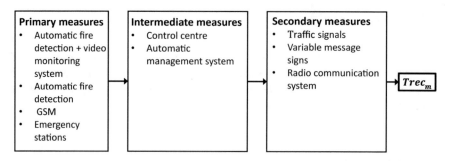

Fig. 8.3 Example of primary, intermediate and secondary measures that influence the recognition time $Trec_m$

The procedure for calculating the four times that make up the pre-movement time $Tpr_{i,j,m,p}$ is described below.

Queue Formation Time

For the generic j-th cell of the i-th lane, th is time is determined as follows:

$$Tfc_{i,j,m,p} = \begin{cases} \frac{(tcl_m+(L_p-Lqcltot_{i,m})/v_i)*xG_{i,j}}{Lqcltot_{i,m}}, & tsat_{i,p} > tcl_m \\ \frac{tsat_{i,p}*xG_{i,j}}{L_p}, & tsat_{i,p} < tcl_m \end{cases}$$

where $(L_p - Lqcltot_{i,m})/v_i$ is the time that the last vehicle to enter the tunnel takes to place itself in the queue before the tunnel is closed.

Recognition Time

Similarly to what has already been described for calculating the tunnel closing time, tcl_m, in Sect. 4.1, the procedure involves a basic recognition time $Trecbase$ that is increased as the system measures MI_m in the tunnel decrease, combined with coefficients for the three measure classes using the formula:

$$Trec_m = Trecbase * \prod_{m=1}^{Msec} Csec_m$$

It is hypothesised that the presence of system measures influences the recognition process of the accidental event by the user in the queue.

Figure 8.3 gives an example of the measures MI_m that influence $Trec_m$.

Figure 8.4 illustrates a numerical example for calculating $Trec_m$ that refers to the infrastructure measures, equipment and management procedures MI_m of Fig. 8.3.

Response Time

With a procedure similar to the one described for the recognition time, a response time is defined considering that $Trbase$ can be increased as the available measures

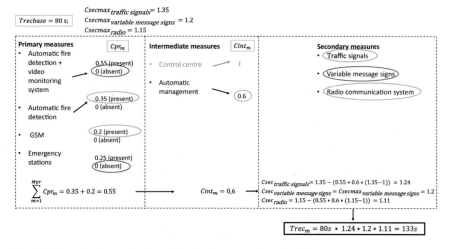

Fig. 8.4 Numerical example for calculating the recognition time $Trec_m$

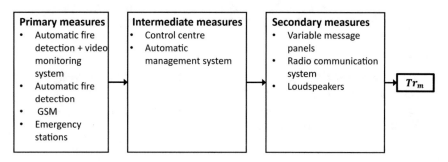

Fig. 8.5 Example of primary, intermediate and secondary measures that influence the recognition time Tr_m

decrease through multiplication coefficients for the three classes of infrastructure measures, equipment and management procedures MI_m with the following formula:

$$Tr_m = Trbase * \prod_{m=1}^{Msec} Csec_m$$

Figure 8.5 gives an example of the measures MI_m that influence Tr_m.

Figure 8.6 illustrates a numerical example for calculating Tr_m referring to the system and infrastructure measures MI_m of Fig. 8.5.

Time to exit the vehicle

The time taken by the user to exit the vehicle is given by the weighted average of the time to exit a light vehicle, a heavy vehicle and a bus according to the composition of the traffic in the *i-th* lane, and is determined as follows:

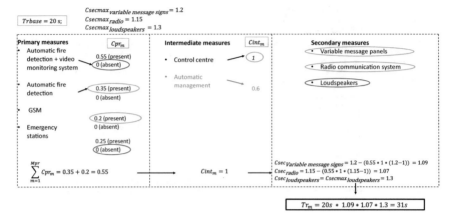

Fig. 8.6 Numerical example for calculating the recognition time Tr_m

$$Tuvehic_i = \frac{\%LV_i * TuLV + \%HV_i * TuHV + \%BUS_i * TuBUS}{100\%}$$

where:

- $TuLV$, $TuHV$, $TuBUS$ are the times required to exit from a light vehicle, a heavy vehicle and a bus, respectively;
- $\%LV_i$, $\%HV_i$, $\%BUS_i$ are, respectively, the percentage of light vehicles, heavy vehicles and buses in the i-th lane.

8.2.2   Movement Time

The user movement time associated with a cell j is given by the following formula:

$$Tmov_{i,j,m} = \frac{dcell_{i,j}}{v_m}$$

where:

- $dcell_{i,j}$ is the distance covered to reach the closest emergency exit defined previously;
- v_m is the user movement speed and is influenced by the infrastructure measures, equipment and management procedures MI_m.

Also in this case a basic movement speed $vbase$ is considered, which is increased as the available measures, combining it with the coefficients associated with the three measurement classes; v_m can be determined using the following formula:

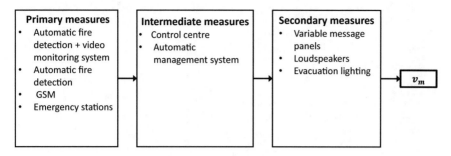

Fig. 8.7 Example of primary, intermediate and secondary measures that influence the user movement speed v_m

$$v_m = vbase * \left(\prod_{m=1}^{Msec} Csec_m \right) * c_s$$

where:

- c_s is a coefficient that considers the effects of the s-th accidental scenario on the movement speed of the user in the tunnel, for example a fire of 15 MW will have less influence on v_m than a 150 MW fire. Numerical examples of c_s relative to three different fire HRR are shown below:

$$c_{15MW} = 1,3; \quad c_{50MW} = 1; \quad c_{150MW} = 0,7$$

- $Csec_m$, unlike what was carried out previously for the tunnel closing, recognition and response times, it can be determined for each infrastructure measures, equipment and management procedures MI_m in two ways:

Case 1

If the generic secondary measure MI_m is not present, $Csec_m = 1$.

In the example in Fig. 8.7, if the variable message panel (VMP) and loudspeakers were absent, the result would be:

$$v_m = vbase * 1 * 1 = vbase$$

In this case the movement speed v_m would become the lowest value, which corresponds to *vbase*.

Case 2

If the generic secondary measure MI_m is present, the intermediate and primary measures are considered.

The coefficients of these measures are the same as those defined previously, and $Csec_m$ is determined as follows:

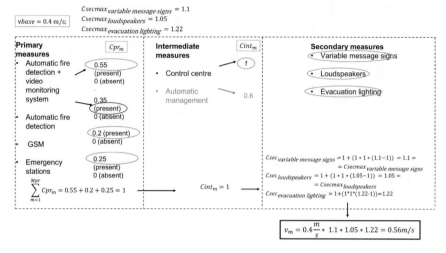

Fig. 8.8 Calculation of v_m with all the measures present

$$Csec_m = 1 + \left(\sum_{m=1}^{Mpr} Cpr_m * Cint_m * (Csecmax_m - 1) \right)$$

Figure 8.7 shows an example of the measures MI_m that influence v_m.

With reference to Figs. 8.7 and 8.8 shows a numerical example for calculating v_m. It can be seen that for both secondary measures $Csec_m = Csecmax_m$, therefore the highest value of v_m is obtained.

8.3 Procedure for the Evaluation of Users' Tenability

To verify the survival of those users who follow the egress process, it is necessary to define $Tpmax$ as the visibility threshold of the zone, included in $z2_s$, involved by the effects of an accidental scenario.

By defining $Tppr_{i,j,m,p,s}$ as the permanence time of users in a cell j located in $z2_s$ during the pre-movement phase, they can survive if $Tppr_{i,j,m,p,s} < Tpmax$; in the contrary case, the users are considered as being dead.

The model involves three types of verification:

- Verification of the user pre-movement time;
- First verification during the egress process;
- Subsequent verifications during the egress process towards the closest emergency exit.

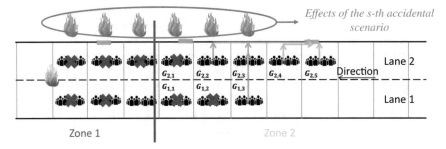

Fig. 8.9 Comparison between the permanence time associated with the pre-movement time and the progress time of the scenario effects

8.3.1 Comparison Between ASET and Pre-movement Time

The first verification consists of comparing the pre-movement time $Tpr_{i,j,m,p}$, with the ASET given by each accidental scenario Tf_s for each j-th cell situated in the i-th lane. In this manner, it is possible to determine the permanence time $Tppr_{i,j,m,p,s}$ as follows:

$$Tppr_{i,j,m,p,s} = \begin{cases} 0 & \text{if } Tpr_{i,j,m,p} < Tf_s(G_{i,j}) \\ Tpr_{i,j,m,p} - Tf_s(G_{i,j}) & \text{if } Tpr_{i,j,m,p} \geq Tf_s(G_{i,j}) \end{cases}$$

Figure 8.9 gives an example of some cases that demonstrate what has just been presented:

- The users in cells $G_{1,1}$, $G_{2,1}$ and $G_{1,2}$, having started the evacuation process too late, accumulating a permanence time $Tppr_{i,j,m,p,s}$ inside the critical zone such that $Tppr_{i,j,m,p,s} \geq Tp\,max$, therefore they are considered to be dead;
- The users in cells $G_{2,2}$, $G_{1,3}$ and $G_{2,3}$ implemented the evacuation process by heading towards the closest emergency exit even though the scenario effects reached them because they were in the critical zone for a time $Tppr_{i,j,m,p,s} < Tp\,max$;
- The users in cells $G_{2,4}$ and $G_{2,5}$ started the evacuation process towards the emergency exit before the scenario effects reached them, therefore $Tppr_{i,j,m,p,s} = 0$.

8.3.2 Dynamic Comparison Between Tenability and Users Egress Times

The following verification is carried out during user movement (evacuation process), after having discretized zone 2 of the tunnel with a constant step P.

In a similar manner to what was done to verify the user pre-movement time, $Tper_{i,j,m,s}$ is defined as the permanence time in the critical zone during the user

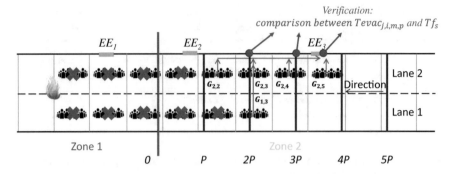

Fig. 8.10 Discretization of $z2$ at constant step P for the comparison between $Tevac_{i,j,m,p}$ and Tf_s

evacuation phase and it is accepted that users can survive in the critical zone if $Tper_{i,j,m,s} < Tpmax$; in the contrary case, the users are considered dead.

For each cell belonging to $z2_s$, the evacuation time $Tevac_{i,j,m,p}$ is compared with the progress time of the effects of each accidental scenario Tf_s. This comparison is carried out at the first step, $1P$, as follows:

$$Tper_{i,j,m,p,s}(1P) = \begin{cases} Tppr_{i,j,m,p,s} & \text{if } Tevac_{i,j,m,p} < Tf_s \\ Tppr_{i,j,m,p,s} + \left(\frac{P}{v_m}\right) & \text{if } Tevac_{i,j,m,p} \geq Tf_s \end{cases}$$

If $Tevac_{i,j,m,p} < Tf_s$, the permanence time associated with evacuation $Tper_{i,j,m,p,s}$ in the critical zone does not increase compared to the permanence time associated with pre-movement $Tppr_{i,j,m,p,s}$; in the opposite case, an additional permanence time, defined as P/v_m, is added.

This operation is carried out at a constant step P for each j-th cell belonging to the i-th lane until the closest emergency exit where the user is hypothesised as being in a safe place.

Figure 8.10 shows an example of discretization of $z2_s$ with step P where it can be seen that:

- for the cell $G_{2,2}$ a comparison is made between $Tevac_{i,j,m,p}$ and Tf_s in correspondence with $2P$, $3P$ and finally at the exit EE_3;
- for cells $G_{1,3}$, $G_{2,3}$ and $G_{2,4}$, $Tevac_{i,j,m,p}$ is compared with Tf_s in correspondence with $3P$ and finally exit EE_3;
- for cell $G_{2,5}$, the comparison between $Tevac_{i,j,m,p}$ and Tf_s is made only at the exit EE_3.

It is necessary to specify that if the value of P is too high, the model would become less precise; it is, in fact, reasonable to assign it with a value similar to that of the length lc of the cell. In this manner, model precision increases because the error committed during verifications at the first step P and near the emergency exit can be considered as negligible, as in these two cases the distance covered is less than P.

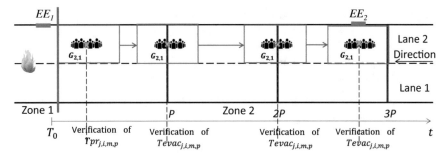

Fig. 8.11 Overall verifications on a generic cell during the evacuation process

Figure 8.11 shows an example of the verifications carried out in order of time t for cell $G_{2,1}$: the first verification involves the user pre-movement time, while the others are carried out at steps P and $2P$ and finally, the last in correspondence with the emergency exit EE_2, comparing the evacuation time.

For the purposes of the verification process carried out on each j-th cell belonging to the i-th lane, the number of fatalities associated with the j-th cell is determined, using the pre-movement time and the evacuation time at step P until the closest emergency exit EE_k, on the basis of the following condition:

$$Ncellfatz2_{i,j,m,p,s} \begin{cases} 0 & \text{if } Tper_{i,j,m,p,s} < Tpmax \\ NC_{i,j,m,p} & \text{if } Tper_{i,j,m,p,s} \geq Tpmax \end{cases}$$

where:

- $Ncellfatz2_{i,j,m,p,s}$ is the number of fatalities associated with the j-th cell located in the i-th lane and belonging to $z2_s$;
- $Tper_{i,j,m,p,s}$ is the permanence time in the critical zone during the user evacuation phase;
- $NC_{i,j,m,p}$ is the number of users in the j-th cell situated in the i-th lane;
- $Tpmax$ represents the visibility threshold of the zone included in $z2_s$ that undergoes the effects of an s-th accidental scenario along the evacuation route.

The total number of fatalities in zone 2 determined by the exceedance of the visibility threshold $Tpmax$ in the area affected by an accident is determined by summing the fatalities of each cell in $z2_s$ where the $Tpmax$ threshold is exceeded:

$$Ntotfatz2_{m,p,s} = \sum_{i=1}^{I} \sum_{j=1}^{Jz2} Ncellfatz2_{i,j,m,p,s}$$

where:

- I is the total number of tunnel lanes;

- $Jz2_i$ is the number of cells in $z2_s$ for the i-th lane;
- $Ntotfatz2_{m,p,s}$ is the total number of fatalities in zone 2, given by the sum of the cells located in $z2_s$ for which $Tper_{i,j,m,p,s} \geq Tp\,max$;
- $Ncellfatz2_{i,j,m,p,s}$ is the number of users associated with the j-th cell who are considered to be dead.

The total number of fatalities inside the tunnel $N_{m,p,s}$ is determined by the sum of $Ntotfatz1_{m,p,s}$ and $Ntotfatz2_{m,p,s}$:

$$N_{m,p,s} = Ntotfatz1_{m,p,s} + Ntotfatz2_{m,p,s}$$

Chapter 9
Calculation of the F-N Curve and the Expected Damage Value

Abstract This chapter describes the implementation process of the F-N curve, which makes possible to represent the societal risk and subsequently verify its acceptance with respect to the ALARP criterion. The F-N curve is evaluated starting from the frequencies of occurrence of the accidental events and the number of fatalities determined by each accidental scenario.

With the aid of the previously described models, it is possible to estimate the total number of fatalities inside the tunnel $N_{m,p,s}$ for each s-th scenario, and each p-th position; this estimate considers the matrix of all the possible combinations c_m of the infrastructure measures, equipment and management procedures.

$N_{m,p,s}$ is then dependant on the $cm - th$ combination of measures, therefore:

$$N_{m,p,s} = N_{cm,p,s}$$

To calculate the F-N curve, it is necessary to determine $NCH_{p,s}$, which represents the number of expected fatalities weighted by the probability H_{cm} of the $cm - th$ combination of measures, for each p-th position and s-th scenario:

$$NCH_{p,s} = \sum_{c_m=1}^{2^{My}} N_{cm,p,s} * H_{cm}$$

Subsequently, the relative frequency of occurrence $Fpos_{p,s}$, determined as shown below, is associated to each value of $NCH_{p,s}$:

$$Fpos_{p,s} = F_s * Ppos_p$$

where:

- F_s is the frequency of occurrence of each s-th accidental scenario;
- $Ppos_p$ is the probability of occurrence of the s-th accidental scenario at the p-th position. The model considers at least 6 different positions, with a probability that

© The Author(s) 2019
F. Borghetti et al., *Road Tunnels*, PoliMI SpringerBriefs,
https://doi.org/10.1007/978-3-030-00569-6_9

Fig. 9.1 Implementation of the F-N curve: sum of the frequencies $Fpos_{p,s}$

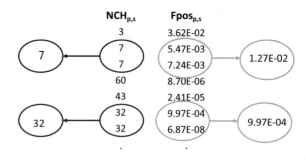

Fig. 9.2 Implementation of the F-N curve: placing the values of $NCH_{p,s}$ in decreasing order

$NCH_{p,s}$	$Fpos_{p,s}$
60	8.70E-06
43	2.41E-05
32	9.97E-04
7	1.27E-02
3	3.62E-02
.	.
.	.
.	.

Fig. 9.3 Implementation of the F-N: determination of $Fpos_{p,s}$ cumulated

$NCH_{p,s}$	$Fpos_{p,s}$	$Fpos_{p,s}$ cumulated	
60	8.70E-06	8.70E-06	
43	2.41E-05	3.28E-05	
32	9.97E-04	1.03E-03	→ 3.28E-05 + 9.97E-04
7	1.27E-02	1.37E-02	
3	3.62E-02	4.99E-02	
.	.	.	
.	.	.	
.	.	.	

can be modified by the analyst. By default, a uniform probability of occurrence is assumed along the tunnel length; for example 0.167 in the case of 6 positions.

If equal values of $NCH_{p,s}$ are present, the relative frequencies $Fpos_{p,s}$, are summed, as shown in the example in Fig. 9.1.

Then the values of $NCH_{p,s}$ are placed in decreasing order, as shown in Fig. 9.2.

Finally, $Fpos_{p,s}$ cumulated is determined from the cumulated sum of $Fpos_{p,s}$, as shown in the example in Fig. 9.3.

The values of $NCH_{p,s}$ represent the x coordinates of the F-N curve, while the values of $Fpos_{p,s}$ cumulated represent the y coordinates as shown in Fig. 9.4.

X		Y
$NCH_{p,s}$	$Fpos_{p,s}$	$Fpos_{p,s}$ cumulated
60	8.70E-06	8.70E-06
43	2.41E-05	3.28E-05
32	9.97E-04	1.03E-03
7	1.27E-02	1.37E-02
3	3.62E-02	4.99E-02
.	.	.
.	.	.
.	.	.

Fig. 9.4 Construction of the F-N curve

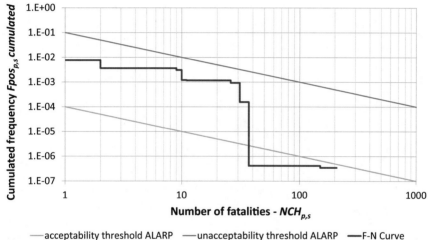

Fig. 9.5 Example of F-N curve

The expected damage value *EDV*, given by the sum of the products between $NCH_{p,s}$ and $Fpos_{p,s}$ *cumulated* for all the points of the F-N curve, is defined as follows:

$$EDV = \sum NCH_{p,s} * Fpos_{p,s} \, cumulated$$

Figure 9.5 shows an example of an F-N curve.

Chapter 10
Model Calibration and Validation

Abstract This chapter describes model calibration in order to verify its consistency. The sensitivity analysis carried out on the parameters and measures considered in the proposed model is described, with the aim of analysing their effect of the shape and position of the F-N curve. In addition, the result of a comparison made using QRAM software and carried out on three tunnels considered to be representative is illustrated.

10.1 Model Calibration

The model was calibrated on a set of tunnels held to be representative, analysing the position and shape of the F-N curve and the EDV in each elaboration.

The parameters on which calibration was based are those tied to the human factors listed below:

- basic recognition time *Trecbase*;
- basic response time *Trbase*;
- time taken to leave the vehicle *Tuvehic*;
- basic movement speed *vbase*;
- threshold of permanence in the damage area of the generic accidental scenario *Tpmax*;
- movement speed coefficients by scenario classes c_s;
- basic tunnel closing time *tclbase*.

To simulate the impact of the measures available on the EDV, so as to reflect the reality in the best possible way, the following coefficients of the primary, intermediate and secondary measures were also calibrated, in order to assign to each of them the correct weight and importance:

- $Cprpres_m$—coefficient of the primary measure present;
- $Cint_m$—coefficient of the intermediate measures;
- $Csecmax_m$—maximum value of the coefficient of the secondary measures.

© The Author(s) 2019
F. Borghetti et al., *Road Tunnels*, PoliMI SpringerBriefs,
https://doi.org/10.1007/978-3-030-00569-6_10

10.2   Sensitivity Analysis

The sensitivity analysis was carried out with the main objective of analysing model behaviour following variations in the system, infrastructure and management parameters and measures. In this manner, it was possible to verify the stability of the proposed model and the importance of parameters and measures for the shape of the F-N curve and the EDV.

The analysis was carried out in a standard configuration tunnel, in which one parameter at a time was varied, halving and doubling its value.

The following parameters were used:

- tunnel ADT;
- basic movement speed *vbase* of the tunnel users;
- basic recognition time *Trecbase*, basic response time *Trbase*, time taken to exit the vehicle *Tuvehic$_i$*;
- survival expectancy threshold *Tpmax* in the area of damage associated with the *s-th* accident;
- initial frequency of occurrence of the initial fire *Finc* and dangerous material spills *Frel* events;
- length of the discretization cell *lc*;
- basic tunnel closing time *tclbase*;
- distance between the emergency exits;
- *%BUS* on total traffic and average occupation coefficient *OBUS*;
- reliability parameter *Pry* connected with measure availability.

In addition, the analysis forecasted the evaluation of each individual system, infrastructure and management measure present in a tunnel of standard configuration. The subsequent paragraphs discuss the results of the analyses carried out on the model parameters and measures.

10.2.1   Tunnel Average Daily Traffic—ADT

The *ADT*, expressed as vehicles/day, is a parameter that influences the queue formation speed inside the tunnel as a consequence of an accidental event, therefore the number of vehicles entering the tunnel before it closes. If the *ADT* is high, the distance between the vehicles that are stopped in the queue and, as a result, the number of users potentially exposed to an accident is greater. If, instead, the *ADT* presents a lower value, the flow of incoming vehicles is smaller, with a greater distance between them, therefore the number of potentially exposed users is lower.

Table 10.1 shows how the *EDV* varies in comparison with the standard configuration, halving and doubling the value of *ADT*. Halving the *ADT* reduces the *EDV* by 21%, because the vehicles in the queue are more spaced and the number of potential users is less; doubling the *ADT* increases the *EDV* by 57%.

Table 10.1 Variation % of the EDV varying the ADT parameter

	Halved parameter	Reference parameter	Doubled parameter
ADT (veh/day)	6710	13,421	26,842
EDV	5.4×10^{-2}	6.8×10^{-2}	1.07×10^{-1}
Variation %	-21%	–	$+57\%$

Table 10.2 Variation % of the EDV varying the *vbase* parameter

	Halved parameter	Reference parameter	Doubled parameter
vbase (m/s)	0.2	0.4	0.8
EDV	1.34×10^{-1}	6.8×10^{-2}	4.65×10^{-2}
Variation %	$+97\%$	–	-32%

10.2.2 Reference Movement Speed—vbase of the Tunnel Users

The basic movement speed *vbase* of the tunnel users is a parameter that influences the evacuation process, because the evacuation time of the individual user towards the closest emergency exit depends on it.

Table 10.2 shows how the *EDV* varies in comparison with the standard configuration value, halving and doubling *vbase*.

When the movement speed is halved, the *EDV* is subjected to an important increase of 97%; this is reasonable because the parameter acts directly on the user evacuation process, and users could remain for an extended period of time in the smoke. Doubling this parameter decreases the *EDV* by 32%.

10.2.3 Reference Recognition and Response Times, and Time Taken to Exit the Vehicle

These three parameters influence the pre-movement time of the users in the tunnel. In particular, the sum of *Trecbase*, *Trbase* and *Tuvehic*$_i$ represents the time that passes from when the vehicle stops in the queue until when the user leaves it and begins the evacuation process towards the emergency exit. The greater these three parameters, the greater the probability that the harmful effects of the accident reach the user before he has left the vehicle.

Table 10.3 demonstrates how the *EDV* varies in comparison with its value in the standard configuration, halving and doubling *Trecbase*, *Trbase* and *Tuvehic*$_i$.

It can be seen that by doubling these parameters, the *EDV* increases by 557%. This high value is justified by the fact that with a much greater pre-movement time value, the majority of users either cannot implement the evacuation process or they

Table 10.3 Variation % of the *EDV* varying the *Trecbase*, *Trbase* and *Tuvehic$_i$* parameters that make up the pre-movement time

	Halved parameters	Reference parameters	Doubled parameters
Trecbase (s)	42.5	85	170
Trbase (s)	12.5	25	50
Tuvehic$_i$ (s)	2.5 for HV, 10 for LV, 120 for BUS	5 for HV, 20 for LV, 240 for BUS	10 for HV, 40 for LV, 480 for BUS
EDV	5.24×10^{-2}	6.8×10^{-2}	4.47×10^{-1}
Variation %	−23%	–	+557%

Table 10.4 Variation % of the EDV varying the *Tpmax* parameter

	Halved parameter	Reference parameter	Doubled parameter
Tpmax (s)	150	300	600
EDV	1.72×10^{-1}	6.8×10^{-2}	4.17×10^{-2}
Variation %	+153%	–	−39%

begin evacuating after being in the damage area for too long: in both cases they cannot save themselves.

Instead, halving the three examined parameters reduces the *EDV* by 23%.

10.2.4   Tenability Threshold Temperature—Tpmax

This parameter defines how long a person can resist in the damage area of an accidental scenario before considered as being dead. It is reasonable to expect that reducing the value of *Tpmax* leads to more fatalities and, as a result, a higher *EDV*, and vice versa.

Table 10.4 shows how the *EDV* varies in relation to the standard configuration value, halving and doubling *Tpmax*.

Halving this parameter makes the *EDV* increase by 153%, while doubling it causes the *EDV* to decrease by 39%.

10.2.5   Frequency of Occurrence of Fires and Dangerous Goods Release

The frequencies of the initial events *Ffire* and *Frel*, expressed in annual events, directly act on the *EDV* value because it is obtained by multiplying the frequencies of each accidental scenario by the number of fatalities. If the initial frequencies

Table 10.5 Variation % of the EDV varying the *Ffire* and *Frel* parameters

	Halved parameters	Reference parameters	Doubled parameters
F_{fire}	7.85×10^{-2}	1.57×10^{-1}	3.14×10^{-1}
F_{rel}	2.07×10^{-3}	4.13×10^{-3}	8.26×10^{-3}
EDV	3.4×10^{-2}	6.8×10^{-2}	1.36×10^{-1}
Variation %	-50%	–	$+100\%$

Table 10.6 Variation % of the EDV varying the *lc* parameter

	Halved parameter	Reference parameter	Doubled parameter
lc (m)	5	10	20
EDV	7.02×10^{-2}	6.8×10^{-2}	7.08×10^{-2}
Variation %	$+3\%$	–	$+4\%$

are doubled, it is reasonable to expect to double the *EDV* obtained in the standard configuration and vice versa.

Table 10.5 shows how the *EDV* varies as to the standard configuration, halving and doubling *Ffire* and *Frel* respectively.

In agreement with expectations, doubling the *Ffire* and *Frel* frequencies causes the *EDV* to double, while halving them causes it to halve.

10.2.6 Effect of the Length of the Discretization Cell

The *lc* parameter influences the distribution of users in the queue inside the tunnel. From the sensitivity analysis, it is reasonable to expect *EDV* values that are similar to each other, because a variation in *lc* changes the distribution of the potentially involved users but not their total number.

Table 10.6 shows how the *EDV* varies as to the value of the standard configuration, halving and doubling *lc*.

Halving *lc* increases the *EDV* by 3%, while doubling *lc* causes the *EDV* to increase by 4%. This result shows that, if the cell size is sufficienlty small (below about 20 m), the model predictions are almost insensitive to the size of the cells which used in the discretization of the queue and in the modeling of the egress.

It can therefore be confirmed that varying *lc* to values of 5 and 20 m has no practical effect in terms of overall damage, therefore the analyst can discretize the queue length with the *lc* value that is held to be most suitable.

An important aspect to consider, however, is the calculation time associated with the cell dimension: as the length of the cell *lc* decreases, the calculation times increase because the longer the queue in the *i-th* lane becomes, the more cells have to be analysed.

Table 10.7   Case 1: variation % of the EDV varying the tclbase parameter

	Halved parameter	Reference parameter	Doubled parameter
tclbase (s)	75	150	300
EDV	6.8×10^{-2}	6.8×10^{-2}	6.8×10^{-2}
Variation %	–	–	–

10.2.7   Reference Tunnel Closing Time

Two cases hypothesizing the basic tunnel closing time are given below.

Case 1: variation of tclbase with respect to the reference value

The closing time is a parameter that influences the length of the queue inside the tunnel after an initial event: the shorter the closing time the shorter the queue created and, as a result, also the number of potential users involved.

Table 10.7 shows the *EDV* in the standard configuration, and in the configurations with *tchlase* halved and doubled.

An apparently anomalous result can be observed because the EDV does not change. A careful analysis clarified that this is caused by the fact that the variation in the queue length, caused by different *tclbase* values, is not influenced by the progress of the negative effects produced by the accidental scenarios examined, as demonstrated in Fig. 10.1.

The EDV does not change because the number of fatalities remains the same.

To highlight a variation in terms of *EDV*, the queue length in the three cases must be of the same size as the maximum progress of the effects, as demonstrated in Fig. 10.2.

In this case, it can be seen that the smaller the queue is, the lower the potential fatalities in the tunnel are.

Case 2: variation of tclbase with respect to the reference value with the progress effects speed increased by a factor of ten

In case 2, the effect of *tclbase* was analysed considering only the fires scenarios (8, 15, 50, 100, 150 MW) and adopting a progress speed ten times greater than that of case 1. Even though this is not meaningful from a physical standpoint, this analysis is useful for verifying that the model responds correctly as *tclbase* varies. Table 10.8 presents *EDV* in the standard configuration, and in the configurations with *tclbase* halved and doubled.

It can be seen that in the standard configuration the *EDV* of case 2 (3.7×10^{-1}) is about four times higher than the *EDV* of the standard configuration of case 1 (6.8×10^{-2}).

In case 2, in addition, halving *tclbase* the *EDV* decreases by 30%, doubling it the *EDV* increases by 31%.

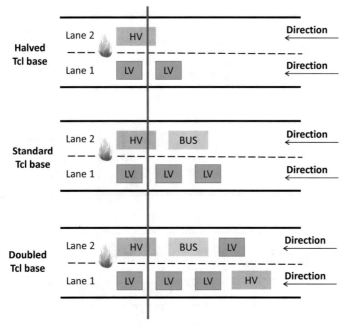

Fig. 10.1 Comparison between the variation in queue length and the maximum progress of the effects, varying *tclbase* (without varying the exposed users)

10.2.8 Distance Between the Emergency Exits

Two cases assuming different distance values between the tunnel emergency exits are given below.

Case 1: variation in the distance between the emergency exits with respect to the reference value

Case 1 analyses the sensitivity of the distance between the emergency exits considering the positions of the accidental event PO_p represented in Table 10.9. They refer to the percentage of the total tunnel length $Ltot$.

The distance between the emergency exits is a geometric parameter of the tunnel which is important for the proposed model. Decreasing this distance increases the number of emergency exits: it is therefore easier for the user to reach an emergency exit.

There can, however, be cases in which an accident occurs nearby an emergency exit: the users who proceed towards it by walking in the direction of the accidental event within the distance *y* could experience the harmful or lethal effects of an accidental scenario. This situation depends on the generic position of the accidental

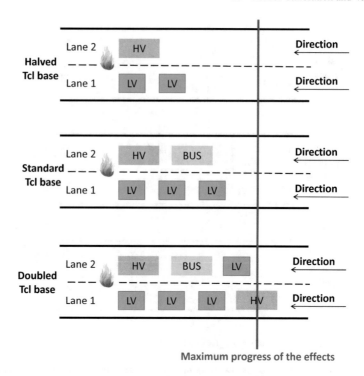

Maximum progress of the effects

Fig. 10.2 Comparison between the variation in the queue length and the maximum progress of the effects, varying *tclbase* (with variation of the exposed users)

Table 10.8 Case 2: variation % of the EDV varying the *tclbase* parameter. Speed of the effects from fire ten times higher than case 1

	Halved parameter	Reference parameter	Doubled parameter
tclbase (s)	75	150	300
EDV	2.6×10^{-1}	3.7×10^{-1}	4.85×10^{-1}
Variation %	-30%	–	$+31\%$

Table 10.9 Case 1: position of the initial event *Pop* as a percentage compared to the tunnel length (case 1)

Positions of the initial event PO_p as a percentage compared to the tunnel length $Ltot$ (case 1)
$PO_1 = 100\%$
$PO_2 = 87\%$
$PO_3 = 75\%$
$PO_4 = 63\%$
$PO_5 = 27\%$
$PO_6 = 15\%$

Table 10.10 Case 1: variation % of the EDV varying the distance between the emergency exits

	Halved parameter	Reference parameter	Doubled parameter
Distance between exits (m)	250	500	1000
EDV	5.12×10^{-2}	6.47×10^{-2}	7.37×10^{-2}
Variation %	-21%	$-$	$+14\%$

Table 10.11 Case 1: position of the closest emergency exit considering the standard distance between exits (500 m) and the halved distance (250 m)

Position of *Pop* (%)	Position of *Pop* (m)	Distance between the position of the initial event PO_p and the first emergency exit	
		Reference parameter (500 m between each emergency exit) (m)	Reference parameter (250 m between each emergency exit) (m)
$PO_1 = 100$	2108	108	108
$PO_2 = 87$	1834	333	83
$PO_3 = 75$	1581	81	81
$PO_4 = 63$	1328	328	80
$PO_5 = 27$	569	69	69
$PO_6 = 15$	316	316	70

event PO_p compared to the position of the closest emergency exit and is governed by randomness: consequently, in this case, the results of the model will be variable according to the tunnel geometry.

Having made this brief introduction, Table 10.10 shows how the *EDV* varies as to the value of the standard configuration, halving or doubling the distance between the emergency exits.

When the distance between the emergency exits was halved, the *EDV* decreased by 21%. Table 10.11 shows the distance between the emergency exit and the position of the accidental event. Referring to position 2, it can be seen that halving the distance, the first exit is 83 m away, inside the damage area of the accidental scenarios; the same thing occurs for the positions 4 and 6. Thanks to the reduction of the distance to reach the closest emergency exit, the evacuating users can reach a safe place more quickly, therefore the number of fatalities decreases and as a result also the *EDV*.

On the contrary, doubling the distance between the emergency exits (1000 m) increased the EDV by 14%. By observing Table 10.12 it can be seen that for position 3 and position 5 the closest emergency exit is located respectively at 581 and 569 m, therefore it is reasonable to expect an increase in the fatalities and, consequently, in the *EDV*.

Table 10.12 Case 2: position of the closest emergency exit considering the standard distance between exits (500 m) and the doubled distance (1000 m)

Position of PO_p (%)	Position of Pop (m)	Distance between the position of the initial event PO_p and the first emergency exit	
		Reference parameter (500 m between emergency exits) (m)	Reference parameter (1000 m between emergency exits) (m)
$PO_1 = 100$	2108	108	108
$PO_2 = 87$	1834	333	833
$PO_3 = 75$	1581	81	581
$PO_4 = 63$	1328	328	328
$PO_5 = 27$	569	69	569
$PO_6 = 15$	316	316	316

Table 10.13 Comparison of the EDV between case 1 with the emergency ventilation system present and case 2 with no emergency ventilation system. Distance between the emergency exits equal to 500 m

	Reference parameter case 1: emergency ventilation system present	Reference parameter case 2: no emergency ventilation system
EDV	6.47×10^{-2}	1.08×10^{-1}
Variation %	–	+67%

Case 2: variation in the distance between the emergency exits compared with the reference configuration, with the same POp positions of case 1 and no emergency ventilation

In case 2, the analysis was carried out considering that there was no emergency ventilation system. The positions of the accidental events are the same as those evaluated in case 1.

Emergency ventilation is a technical measure that strongly affects the progress of the accident effects. With no ventilation, the effects spread more rapidly and for a longer time along the tunnel, making the evacuation process of the users more critical.

Table 10.13 shows the comparison between the *EDV* of the standard configuration of case 1 with the presence of an emergency ventilation system and the *EDV* of the standard configuration of case 2, with no emergency ventilation. The lack of emergency ventilation causes the *EDV* in case 2 to increase by 67%.

Finally, the variation of the *EDV* with no emergency ventilation is analysed, varying the value of the distance between the emergency exits. Table 10.14 shows that halving the distance decreases the *EDV* by 35%, while doubling the distance increases the *EDV* by 25%. The % variation is greater than that of case 1: it can be seen that without emergency ventilation the presence of an emergency exit in the damage area has a more favourable impact on the *EDV* that in the case with ventilation.

Table 10.14 Case 2: % variation of the EDV varying the distance between the emergency exits. No emergency ventilation system

	Halved parameter	Reference parameter	Doubled parameter
Distance between exits (m)	250	500	1000
EDV	7.02×10^{-2}	1.08×10^{-1}	1.35×10^{-1}
Variation %	-35%	–	$+25\%$

10.2.9 Number of Positions

This analysis is performed with the aim of assessing that the number of fire positions used in the model PO_p ($=6$) is sufficient to represent the effect of the distribution of accidental events along a generic tunnel. For this reason, a sensitivity analysis was performed to investigate the effects of varying the number of fire positions PO_p on model results:

- 1 position
- 3 positions
- 6 positions (reference value used in the model)
- 12 positions
- 18 positions

This sensitivity analysis was performed for 4 tunnels, characterized by different length, and infrastructure measures, equipment and management procedures. For each tunnel tube, two configurations were considered, as indicated in Table 10.15:

- Real configuration, which considers the real position of the emergency exits (bypasses) and the equipment and infrastructure measures actually available inside the tunnel.
- Virtual configuration, which assumes that the position of the emergency exits (bypasses) and the equipment and infrastructure measures available inside the tunnel are consistent with the law (European Directive 2004/54/EC), which requires a bypass every 500 m for tunnels longer than 1000 m).

The aim of this analysis is not to compare the two configurations, but rather to assess the sensitivity of model results to the number of initial positions. The aim is to adopt a number of initial positions sufficiently large to guarantee that the predictions in terms of F-N curve are not affected by the assumed positions. It is in fact important to notice that the initial positions of fire events cannot be predicted, since they can occur in any arbitrary positions along the tunnel. In this analysis a uniform distribution of probability is assumed along the positions.

Tunnel 1

Figure 10.3 shows the position of initial events inside tunnel 1. Since the tunnel is short, bypasses are not present. Initial event is uniformly distributed along the tunnel length, both in terms of position and probability.

Table 10.15 Tunnel length and emergency exit positions along the tunnel for the real and virtual configuration

Tunnel length and bypass positions along the tunnel (m)							
Tunnel 1 L = 805 m		Tunnel 2 L = 2108 m		Tunnel 3 L = 4547 m		Tunnel 4 L = 10,121 m	
Configuration		Configuration		Configuration		Configuration	
Real	Virtual	Real	Virtual	Real	Virtual	Real	Virtual
–	–	518	500	539	500	566	500
		1043	1000	837	1000	1075	1000
		1594	1500	1080	1500	1791	1500
			2000	1388	2000	2441	2000
				1680	2500	3123	2500
				1980	3000	3835	3000
				2263	3500	4479	3500
				2288	4000	5056	4000
				2577	4500	5607	4500
				2877		6026	5000
				3177		6578	5500
				3475		7080	6000
				3791		7104	6500
				4077		7590	7000
						7938	7500
						8227	8000
						8663	8500
						9366	9000
							9500
							10,000

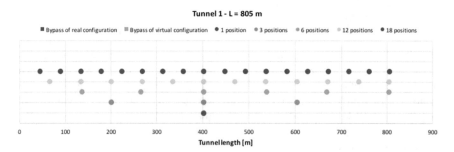

Fig. 10.3 Tunnel 1: accident positions and bypass locations for the real and virtual configuration

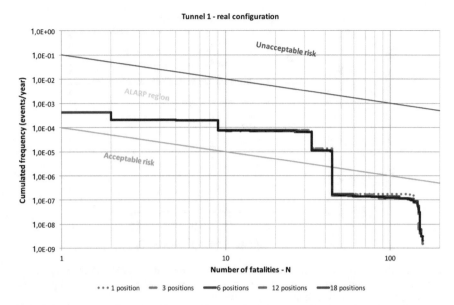

Fig. 10.4 Tunnel 1: effect of accident positions on the F-N curves for the real configuration

Figure 10.4 shows a comparison of the F-N curves calculated for tunnel 1 assuming different positions for the initial events. It is interesting to observe that the 5 curves almost overlap, meaning that the model results are not sensitive to the number of initial positions for this tunnel.

To better highlight this point, Fig. 10.5 shows the negligible effect of a different number of initial positions on the EDV value for tunnel 1 for the real configuration. The dashed line is the average EDV value calculated using 6, 12 and 18 positions. The dotted lines represent a change of ±10% of the average EDV value.

It is possible to observe that the EDV value predicted using 6 positions is very close to the EDV value calculated using a larger number of positions (12, 18).

Figures 10.6 and 10.7 show the F-N curves and the EDV values for the virtual configuration of tunnel 1. Also for the virtual configuration the sensitivity of model results to the number of initial position is negligible.

Tunnel 2

Figure 10.8 shows the location of the bypasses in the tunnel and of the positions of the initial events for Tunnel 2 in both real and virtual configuration.

Figure 10.9 shows that for this longer tunnel, with presence of bypasses, the sensitivity to the number of initial positions is more important than for tunnel 1. However, the shape of the different F-N curves is similar. More importantly, Fig. 10.10 confirms that there is a scarce sensitivity to the number of initial positions when the EDV value is considered, if at least 6 positions are considered. On the contrary, assuming 1 or 3 positions will affect the results with deviations larger than 10%.

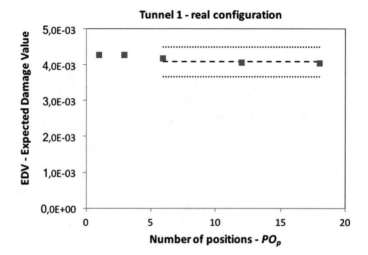

Fig. 10.5 Tunnel 1: EDV as a function of number of accident positions for the real configuration. The dashed line represents the average value of EDV for scenarios with 6, 12 and 18 positions. Dotted lines represent ±10% of the average EDV value

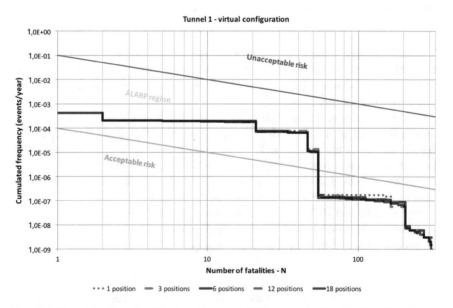

Fig. 10.6 Tunnel 1: effect of accident positions on the F-N curves for the virtual configuration

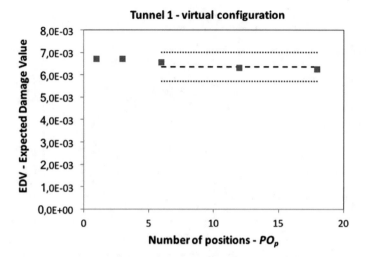

Fig. 10.7 Tunnel 1: EDV as a function of number of accident positions for the virtual configuration. The dashed line represents the average value of EDV for scenarios with 6, 12 and 18 positions. Dotted lines represent ±10% of the average EDV value

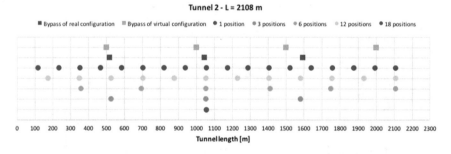

Fig. 10.8 Tunnel 2: accident positions and bypass locations for the real and virtual configuration

Figure 10.11 show the analogous comparison for Tunnel 2 assuming the virtual configuration. It is possible to observe that the F-N curves tend to converge and overlap when the number of initial positions is increased.

Figure 10.12 confirms this observation in terms of the EDV value, showing that also in this case 6 positions are sufficient to ensure grid-independent results.

Tunnel 3

Figure 10.13 describes the geometry of Tunnel 3 in terms of bypasses and location of the initial positions for the real and virtual configurations. Since the tunnel is quite long, a large number of bypasses is present.

Figure 10.14 shows the effect of the different number of positions on the F-N curves for the real configuration of Tunnel 3. Only the F-N curve calculated assuming only 1 position is significantly different form the others.

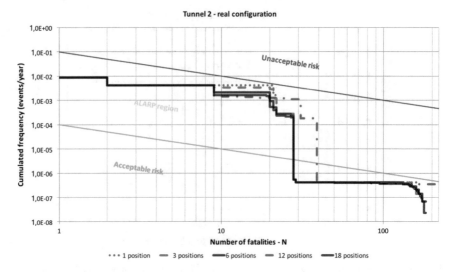

Fig. 10.9 Tunnel 2: effect of accident positions on the F-N curves for the real configuration

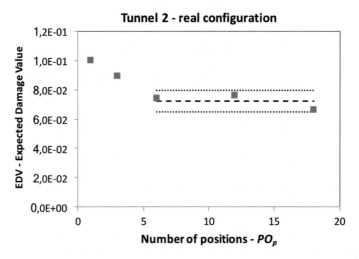

Fig. 10.10 Tunnel 2: EDV as a function of number of accident positions for the real configuration. The dashed line represents the average value of EDV for scenarios with 6, 12 and 18 positions. Dotted lines represent ±10% of the average EDV value

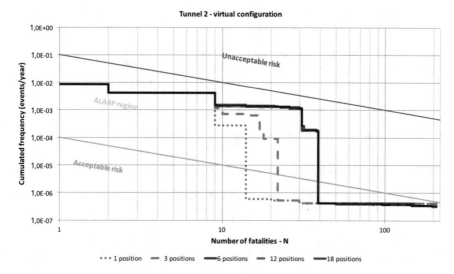

Fig. 10.11 Tunnel 2: effect of accident positions on the F-N curves for the virtual configuration

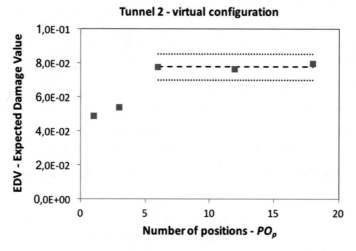

Fig. 10.12 Tunnel 2: EDV as a function of number of accident positions for the virtual configuration. The dashed line represents the average value of EDV for scenarios with 6, 12 and 18 positions. Dotted lines represent ±10% of the average EDV value

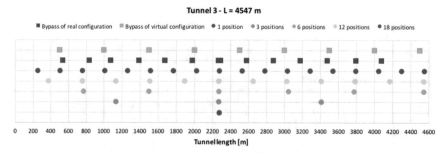

Fig. 10.13   Tunnel 3: accident positions and bypass locations for the real and virtual configuration

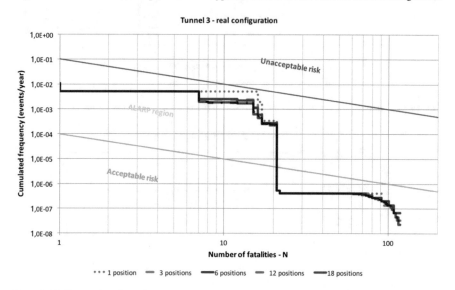

Fig. 10.14   Tunnel 3: effect of accident positions on the F-N curves for the real configuration

Similarly, Fig. 10.15 confirms this result in terms of EDV value. In this case a minimum of 3 positions is sufficient to ensure that model results are within ±10% of the average EDV value.

Figures 10.16 and 10.17 show a similar effect in the case of the virtual configuration of Tunnel 3. The results are similar but not the same of the real configuration mostly because of the different position of the bypasses. However, also in this case a number of initial position equal to 3 is sufficient to achieve a reasonably stable EDV value.

Tunnel 4

Finally, Fig. 10.18 shows the complexity of Tunnel 4, which is very long and characterized by a large number of bypasses. Bypasses are irregularly distributed in the

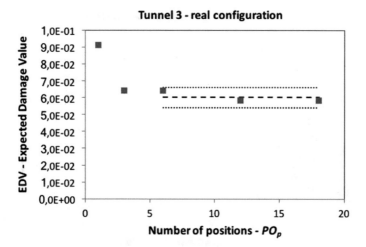

Fig. 10.15 Tunnel 3: EDV as a function of number of accident positions for the real configuration. The dashed line represents the average value of EDV for scenarios with 6, 12 and 18 positions. Dotted lines represent ±10% of the average EDV value

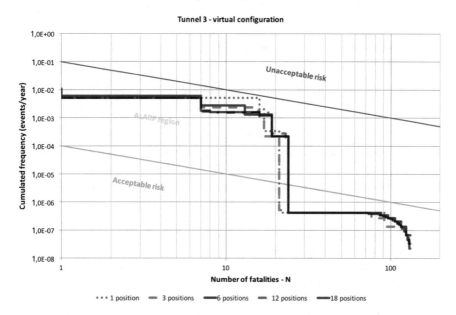

Fig. 10.16 Tunnel 3: effect of accident positions on the F-N curves for the virtual configuration

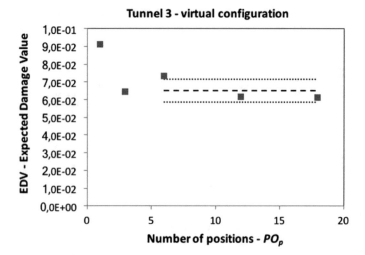

Fig. 10.17 Tunnel 3: EDV as a function of number of accident positions for the virtual configuration. The dashed line represents the average value of EDV for scenarios with 6, 12 and 18 positions. Dotted lines represent ±10% of the average EDV value

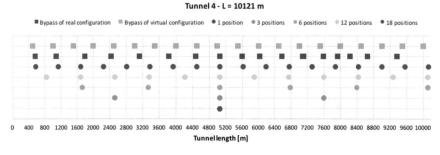

Fig. 10.18 Tunnel 4: accident positions and bypass locations for the real and virtual configuration

real configuration, while are more numerous and equally distributed along the tunnel length in the virtual configuration.

Figure 10.19 show the predicted F-N curves for the real configuration of Tunnel 4. Despite the complexity of the geometry, there is a relatively scarce sensitivity to the number of initial accident positions. In fact, Fig. 10.20 shows that, despite of the large tunnel length, 6 positions are sufficient to ensure a consistent EDV prediction.

Figure 10.21 shows the same comparison for the virtual configuration of Tunnel 4. A deviation can be observed for F-N curves predicted using less than 6 positions. Figure 10.22 shows that also in this case 6 positions are needed, while 3 and 1 positions would predict a significantly smaller value of the EDV.

It is in general important to notice that using a small number of initial positions may affect the predicted F-N curves and EDV values in a different way. In some cases, it would lead to an underestimation of the EDV, as in the case of the virtual

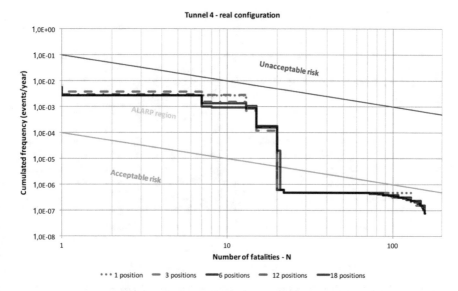

Fig. 10.19 Tunnel 4: effect of accident positions on the F-N curves for the real configuration

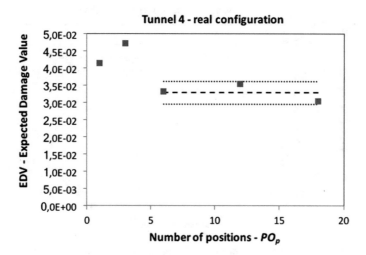

Fig. 10.20 Tunnel 4: EDV as a function of number of accident positions for the real configuration. The dashed line represents the average value of EDV for scenarios with 6, 12 and 18 positions. Dotted lines represent ±10% of the average EDV value

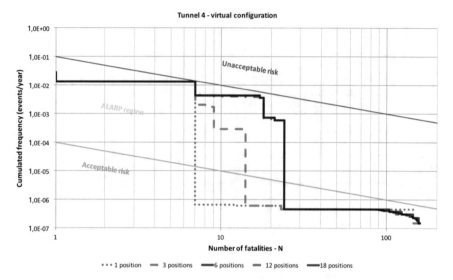

Fig. 10.21   Tunnel 4: effect of accident positions on the F-N curves for the virtual configuration

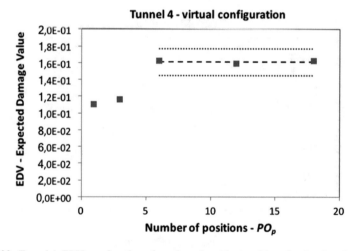

Fig. 10.22   Tunnel 4: EDV as a function of number of accident positions for the virtual configuration. The dashed line represents the average value of EDV for scenarios with 6, 12 and 18 positions. Dotted lines represent ±10% of the average EDV value

Table 10.16 Variation % of the EDV varying $\%_{BUS}$ and O_{BUS}

	Reference parameters	Increased parameters case 1	Increased parameters case 2
$\%_{BUS}$	1%	5%	10%
O_{BUS}	30	50	50
EDV	6.8×10^{-2}	1.46×10^{-1}	2.27×10^{-1}
Variation %	–	+115%	+234%

configuration of the Tunnel 4. In other cases to the opposite deviation, as in the real configuration of Tunnel 4. This result mostly depends on the relative position of the accidental events and the bypasses. If the initial event is placed in a favorable position, the egress of the users is facilitated, while the opposite may occur if it is placed in an unfavorable position. Since the positions of the events cannot be predicted a priori, it is important to use a sufficiently large number of positions to ensure that the results are not affected by this assumption.

The analysis performed in this paragraph, which took into account four different tunnels and different configuration, showed that a minimum of 6 positions should be adopted for risk analysis.

10.2.10   Percentage of Buses on the Total Traffic and Buses Occupancy Rate

In this paragraph, the *EDV* will be evaluated by varying the percentage of buses, $\%_{BUS}$, with respect to the total traffic and the average occupation coefficient of the buses, O_{BUS}.

Table 10.16 compares the *EDV* and its variation in the three cases that were examined. As is reasonable to expect, increasing the $\%_{BUS}$ and the respective occupation coefficient, the *EDV* increases because the number of users inside the tunnel increases; in addition, the time to exit a bus is greater than the time to exit a light or a heavy vehicle.

10.2.11   Reliability Parameter Tied to Measure Availability

The last parameter analysed deals with the reliability of the system, infrastructure and management measures *Pry*. Decreasing the value of this parameter makes the *Prn* (complementary) parameter increase, and as such the probability that the *m-th* system measure is faulty or not available when an initial event occurs increases. Table 10.17 shows how the *EDV* increases as the *Pry* is reduced from the standard (0.95) first to 0.7 and then to 0.5.

Table 10.17 Variation % of the EDV varying PR_y

	Reference parameter	Increased parameter case 1	Increased parameter case 2
Pr_y	0.95	0.7	0.5
(Pr_n)	(0.05)	(0.3)	(0.5)
EDV	6.8×10^{-2}	1.04×10^{-1}	1.28×10^{-1}
Variation %	–	+53%	+89%

10.2.12   Tunnel Infrastructure Measures, Equipment and Management Procedures

In this paragraph, we analyse the variation of the *EDV* considering the contribution of the individual tunnel infrastructure measures, equipment and management procedures, in comparison with the configuration without any measure present.

The following secondary measures are considered, independently from any primary and intermediate measure:

- flammable liquid drainage: influences the event tree related to dangerous goods spills at the question *"relevant spill?"*;
- emergency pedestrian platforms: they influence the user movement speed during the evacuation process;
- road signs: they influence the user movement speed during the evacuation process;
- fire-fighting system: influences the event tree related to fire events and the number of total accidental scenarios;
- water supply: influences the event tree of fire events at the question *"is the fire rapidly extinguished?"*;
- emergency ventilation: influences the propagation of the effects of the accidental scenarios and the user movement speed during evacuation;
- specialised rescue team: influences the event tree of fire events at the question *"is the fire rapidly extinguished?"* and also the user movement speed during evacuation.

The following secondary measures, the operation of which depends on the presence of primary and intermediate measures, are also considered:

- speakers in shelters and at emergency exits: they influence user movement speed during the evacuation process and the response time;
- emergency messages by radio for tunnel users: they influence the realisation and answer times of the users;
- traffic lights and/or arrow-cross panels inside the tunnel: they influence the user recognition time;
- variable message panels inside the tunnel: these influence the recognition time, response time and movement speed of the users during evacuation;
- emergency lighting: this influences the user movement speed during evacuation.

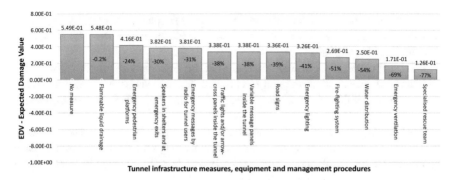

Fig. 10.23 Effect of each individual measure on the EDV, and its variation in comparison with the case without any measure

This analysis considers that all the primary and intermediate measures of the model are available, so permitting effective operation of the secondary measures. In this manner a comparison can be made between all the secondary measures, both independent and those with interdependence, as described in Chap. 4.

The following primary measures were considered:

- automatic event identification through cameras;
- GSM coverage;
- emergency stations.

The intermediate measure is the control centre.

Figure 10.23 shows that the measures which reduce the EDV the most are the specialized rescue team and the emergency ventilation system. Flammable liquid drainage reduces the *EDV* by just 0.2% because this measure acts on the frequencies of occurrence F_s of the accidental scenarios associated with the initial DG release event: as these scenarios are already characterized by a very low frequency, their reduction would not have much effect on the variation in the *EDV*.

10.3 Automation of the Calculation Process

As defined in the previous paragraphs, for an accidental scenario there are 2^{M_y} combinations of analysable measures where M_y represents the number of tunnel infrastructure measures, equipment and management procedures that are present; if the generic tunnel has all 15 measures available, the *c-th* combinations are equal to $2^{15} = 32,768$; considering the 18 accidental scenarios (fires and DG releases) the *c-th* combinations are equal to $18 \cdot 32,768 = 589,824$; each of the 18 accidental scenarios can occur in a different position along the tunnel. Considering the 6 positions PO_p the analysed combinations of measures are equal to $6 * 589,824 = 3,538,944$. Overall, for a single tunnel the model can examine up to 3,538,944 combinations, calculating the number of fatalities and the related frequency of occurrence, thus it provides the F-N curve. If the number of positions is increased, the total number of combinations is even larger.

The computational effort cannot, therefore, be neglected and automation of the calculation process is fundamental to optimize the analysis in terms of time and to reduce the possibility of errors due to the analyst (manual data entry, reading errors etc.).

Considering the quantity of variables and the complexity of the operations to be carried out, the need for automating the model emerged immediately, in order to make data insertion and result acquisition easier. It was, as such, decided to implement it with *VBA—Visual Basic for Applications*.

Visual Basic for Applications is a Visual Basic implementation inside Microsoft applications such as the Office suite or other applications such as AutoCad. In spite of its close bond with Visual Basic, VBA cannot be used to carry out stand-alone applications, but a certain interoperability between applications such as Excel or Word exists however, thanks to the automation. *VBA* is an object programming language: the main objects of this language are subroutine and functions.

The Excel interface was selected to manage the input and output data, while *VBA* was given the duty of automating all the model operations. By writing the model in *VBA* language, it is therefore possible to make the Excel cells read the input data directly. In this way, the large quantity of data can be easily managed. The number of input parameters varies according to the number of emergency exits, the number of lanes and the number of accidental scenarios considered. On average, there are around 140 input parameters, a number that can only be handled with a clear and automated interface.

10.4   Comparison with QRAM Software

In order to verify the reliability of the calibration on a set of representative tunnels, the results were compared using *QRAM* software.

It must be specified that this comparison is carried out follwing a semi-quantitative approach, because the considered scenarios and the selectable tunnel infrastructure measures, equipment and management procedures of the two models are not easy to compare.

The comparison is believed to be useful for verifying predicted trends and the position of the F-N curves.

Three representative tunnels of the one-way motorway type were selected, with the comparison in terms of F-N curve being made for each one.

Table 10.18 respectively shows the main characteristics of tunnel 2, while Fig. 10.24 shows the comparison in terms of F-N curve between the proposed model and the *QRAM* software.

Table 10.18 Main characteristics of tunnel 1

Tunnel 1	
Parameter	Value
Length (m)	910
Section (m²)	54
Number of lanes	2
Distance between the emergency exits (m)	500
ADT (vehicles/day)	5623
Peak time flow when the analysis was carried out (vehicles/h)	562 (10% ADT)
Longitudinal slope (%)	+3.4
Average number of people in a light vehicle (n. people/vehicle)	2
Average number of people in a heavy vehicle (n. people/vehicle)	1.1
Average number of people in a bus (n. people/vehicle)	30
Percentage of light vehicles (%)	85
Percentage of heavy vehicles (%)	14
Percentage of buses (%)	1
Average speed of the light vehicles (km/h)	110
Average speed of the heavy vehicles/buses (km/h)	70

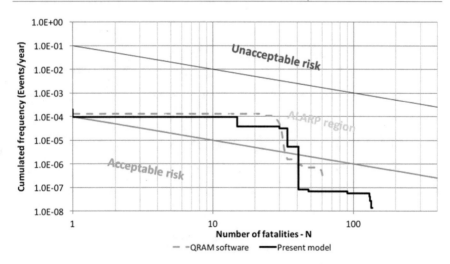

Fig. 10.24 Comparison 1: tunnel length 910 m and ADT 5623 vehicles/day

Table 10.19 Main characteristics of tunnel 2

Tunnel 2	
Parameter	Value
Length (m)	2100
Section (m²)	56
Number of lanes (-)	2
Distance between the emergency exits (m)	500
ADT (vehicles/day)	10,946
Peak time flow when the analysis was carried out (vehicles/h)	1094 (10% ADT)
Longitudinal slope (%)	−1.4
Average number of people in a light vehicle (n. people/vehicle)	2
Average number of people in a heavy vehicle (n. people/vehicle)	1.1
Average number of people in a bus (n. people/vehicle)	30
Percentage of light vehicles (%)	85
Percentage of heavy vehicles (%)	14
Percentage of buses (%)	1
Average speed of the light vehicles (km/h)	110
Average speed of the heavy vehicles/buses (km/h)	70

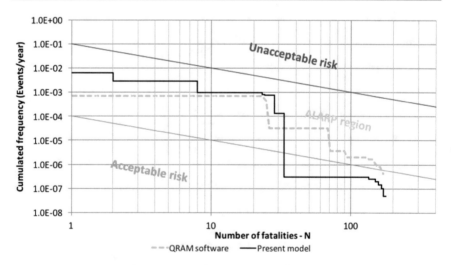

Fig. 10.25 Comparison 2: tunnel length 2100 m and ADT 10,946 vehicles/day

Table 10.19 presents the main characteristics of tunnel 2 and Fig. 10.25 shows the comparison in terms of F-N curve between the proposed model and the QRAM software.

Table 10.20 Main characteristics of tunnel 3

Tunnel 3	
Parameter	Value
Length (m)	4550
Section (m^2)	54
Number of lanes	2
Distance between the emergency exits (m)	500
ADT (vehicles/day)	5053
Peak time flow when the analysis was carried out (vehicles/h)	505 (10% ADT)
Longitudinal slope (%)	−1
Average number of people in a light vehicle (n. people/vehicle)	2
Average number of people in a heavy vehicle (n. people/vehicle)	1.1
Average number of people in a bus (n. people/vehicle)	30
Percentage of light vehicles (%)	80
Percentage of heavy vehicles (%)	19
Percentage of buses (%)	1
Average speed of the light vehicles (km/h)	110
Average speed of the heavy vehicles/buses (km/h)	70

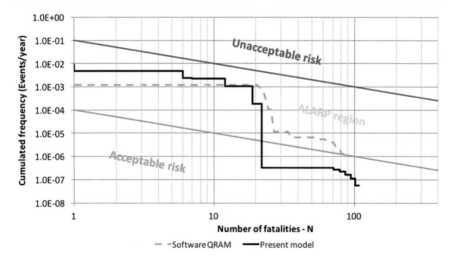

Fig. 10.26 Comparison 3: tunnel length 4550 m and ADT 5053 vehicles/day

Table 10.20 gives the main characteristics of tunnel 2 and Fig. 10.26 presents the comparison in terms of F-N curve between the proposed model and the QRAM software.

Even though it is difficult to make a quantitative comparison between the two models, it is possible to observe that in the three cases investigated there is a consistent correspondence between the two codes, both in terms of F-N curve shape and EDV. In this way, the validity of the proposed model can be confirmed on the basis of the results obtained using the QRAM model, which represents an international reference for tunnel safety. However, the advantage of the present model is related to the possibility to include the effect of numerous infrastructure measures, equipments and management procedures, whose effect cannot be evaluated currently using the QRAM model. Moreover, it is possible to link this risk analysis model to zone models (e.g. CFAST) or even to CFD codes (e.g. FDS) for the prediction of the fire development and smoke fluidynamics. In this way, the model allows to study different configurations for tunnels characterized by complex ventilation strategies, and to perform cost-benefit considerations.

Conclusions

The book illustrates and describes the structure of a quantitative model of risk analysis for road tunnels. The result of the model, in accordance with the European Directive 2004/54/EC and the Italian Legislative Decree 264/2006, provides the F-N curves of societal risk, in other words functions that bind the cumulated frequency of occurrence of different accidental scenarios (F) with the expected consequences in terms of potential victims (N: number of fatalities).

Starting from two initial events, fire and DG—Dangerous Good release, 18 accidental scenarios were defined. The frequency of occurrence of each accidental scenario analysed was obtained using the Event Tree Analysis (ETA) technique.

To estimate the consequences of each accidental scenario, taking as a starting point what is available from scientific literature, three sub-models were implemented: a sub-model for the formation of the queue of vehicles inside the tunnel, a sub-model for the distribution of the users involved, and a sub-model for the simulation of the egress of the users potentially exposed to the consequences of an accidental scenario.

In this way it is possible to simulate the evacuation of those users present in vehicles, comparing the egress times (RSET) needed to reach a safe area (emergency exits or bypasses) with the available time (ASET). This procedure gives an estimate of the number of fatalities.

The length of the queue of vehicles is calculated on the basis of traffic data, the size of the vehicles and the closing time of the tunnel: this offers greater model precision when compared to some currently-existing models that assume the tunnel as being completely full.

The proposed model can simulate each of the 18 accidental scenarios in 18 different positions along the tunnel (this number can be modified by the analyst), considering the impact that the 15 tunnel infrastructure measures, equipment and management procedures have on the users evacuation and on the propagation of the effects of the accidental scenarios.

For each of the 15 measures that can be selected by the analyst, the model can also consider their reliability, hypothesizing that each selected measure can be

© The Author(s) 2019
F. Borghetti et al., *Road Tunnels*, PoliMI SpringerBriefs
https://doi.org/10.1007/978-3-030-00569-6

unavailable at the moment of need; a probabilistic parameter of reliability for each of them was defined in this manner, which involves the analysis of 2^{my} combinations. In each combination, the availability or not of each selected measure is taken into consideration on the basis of its reliability.

Considering 18 accidental scenarios, 18 possible positions and the availability of each measure considered (2^{15}) the model analyses up to 10,616,832 combinations for a single tunnel.

Given the high number of scenarios associated with the sub-models and the large number of input parameters (more than 140 parameters can be inserted or modified by the analyst), the proposed model was implemented in *VBA—Visual Basic for Applications*, using the *Excel* interface for data entry and results reading. As such, it is possible to automate the calculation process, reducing the possibility of errors made by the analyst.

The model was calibrated considering a set of tunnels that represent the motorway environment in terms of length, traffic, geometry, shape, etc., verifying the consistency of the predictions in terms of the position of the F-N curve and the EDV.

A sensitivity analysis was also carried out on the most important model parameters, in order to evaluate their relative weight and importance. It was decided to halve and double the value of each parameter as compared to a reference configuration, in order to verify the variation of the EDV. It emerged from this analysis that the parameter with the largest impact on user evacuation was the pre-movement time, followed by the movement speed and the visibility threshold in the damage area of the accidental scenarios. Some parameters, such as the tunnel closing time and the distance between the emergency exits, only have an impact on the overall damage if placed in relation to other parameters, for example the damage area of the generic accidental scenario or the presence of emergency ventilation.

An additional sensitivity analysis was carried out considering tunnel infrastructure measures, equipment and management procedures. The specialised rescue team and emergency ventilation are the measures that most influence the EDV, while the measures that act on the frequency to have a spill of dangerous material, such as flammable liquid drainage, have an impact that affects the EDV only slightly, because these frequencies of occurrence are small in comparison to those of fire events.

Finally, to validate the model, comparisons were made with the QRAM software developed by PIARC; for three representative tunnels, the F-N curves were evaluated and compared. It must be specified that comparing the two models was particularly difficult for at least two reasons: the scenarios analysed are not exactly comparable and in the QRAM software is not possible to select individually the measures present in the tunnel, in accordance with the European Directive. Even the logic of the two models is quite different. The semi-quantitative comparison that was carried out confirmed the validity and consistency of the proposed model because both trend and position of the F-N curves can be compared in all the three cases that were analysed.

Other elements that could be developed involve the possibility of assigning to each basic parameter of the model, for example the pre-movement time and the user movement speed, a stochastic distribution with respect to an average value, thus obtaining a different base value for each simulation carried out.

Development of the user distribution model inside the tunnel considers the length of each different discretization cell according to the type of vehicle traffic; in this case a high number of scenarios must be evaluated, operating a series of permutations, in order to simulate the possible combinations and reciprocal disposition of the different types of vehicle (light, heavy and buses), through statistical approaches such as the Montecarlo methods.

In addition, instead of concentrating the users in a queue in the centre of the cells into which the tunnel is discretized, it could be considered the possibility to define different reference positions in which to place the users, and estimate subsequently their evacuation paths. In fact, the reference positions could lead to different values according to the type of vehicle: with an abundance of light vehicles the accepted reference position is the central point of the cell, whereas with a prevalence of heavy vehicles the position is anticipated and taken to 1/3 of the cell length (to simulate the fact that the driver's cabin in heavy vehicles is not the physical centre of gravity of the vehicle). Finally in case of an important percentage of buses, the passengers are distributed evenly at two points, at ¼ and ¾ of the cell length (to simulate the presence of more than one door for the users to exit the bus).

In the proposed model, the hypothesis of concentrating the users in the central point of the cell is considered to be reasonable, because using cells of limited length is suggested (around 10 m). In cases of greater cell lengths, it may be convenient to consider a distribution of users similar to the one previously proposed.

Printed in the United States
By Bookmasters